International Perspectives on Age-Friendly Cities

This book brings together recent scholarly work concerned with efforts around the world to transform cities so that they are more age-friendly. Common to all of the initiatives is recognition of the importance of the community environment for the well-being of the rapidly growing numbers of older people. The collection includes chapters that examine the circumstances in which communities currently undertake significant age-friendly initiatives, public-private collaboration in age-friendly initiatives, collaboration across institutional sectors in age-friendly initiatives, policies that facilitate age-friendly developments, and the bases upon which age-friendly initiatives should be evaluated. It will be of interest to scholars in various fields, including urban planning, gerontology, transportation planning, environmental design, and adult education.

Francis G. Caro is a professor emeritus of Gerontology at the University of Massachusetts Boston and the editor of the *Journal of Aging & Social Policy*.

Kelly G. Fitzgerald is a senior scientist and adjunct graduate faculty member at Western Kentucky University's Center for Gerontology.

Routledge Studies in Human Geography

This series provides a forum for innovative, vibrant, and critical debate within Human Geography. Titles will reflect the wealth of research that is taking place in this diverse and ever-expanding field. Contributions will be drawn from the main sub-disciplines and from innovative areas of work that have no particular sub-disciplinary allegiances.

Published:

29 **Rural–Urban Dynamics**
Livelihoods, mobility and markets in African and Asian frontiers
Edited by Jytte Agergaard, Niels Fold and Katherine V. Gough

30 **Spaces of Vernacular Creativity**
Rethinking the cultural economy
Edited by Tim Edensor, Deborah Leslie, Steve Millington and Norma Rantisi

31 **Critical Reflections on Regional Competitiveness**
Gillian Bristow

32 **Governance and Planning of Mega-City Regions**
An international comparative perspective
Edited by Jiang Xu and Anthony G.O. Yeh

33 **Design Economies and the Changing World Economy**
Innovation, production and competitiveness
John Bryson and Grete Rustin

34 **Globalization of Advertising**
Agencies, cities and spaces of creativity
James R. Faulconbridge, Peter J. Taylor, Jonathan V. Beaverstock and Corinne Nativel

35 **Cities and Low Carbon Transitions**
Edited by Harriet Bulkeley, Vanesa Castán Broto, Mike Hodson and Simon Marvin

36 **Globalization, Modernity and the City**
John Rennie Short

37 **Climate Change and the Crisis of Capitalism**
A chance to reclaim self, society and nature
Edited by Mark Pelling, David Manual Navarette and Michael Redclift

38 **New Economic Spaces in Asian Cities**
From industrial restructuring to the cultural turn
Edited by Peter W. Daniels, Kong Chong Ho and Thomas A. Hutton

39 **Landscape and the Ideology of Nature in Exurbia**
Green sprawl
Edited by Kirsten Valentine Cadieux and Laura Taylor

40 **Cities, Regions and Flows**
Edited by Peter V. Hall and Markus Hesse

41 **The Politics of Urban Cultural Policy**
Global perspectives
Edited by Carl Grodach and Daniel Silver

42 **Ecologies and Politics of Health**
Edited by Brian King and Kelley Crews

43 **Producer Services in China**
Economic and urban development
Edited by Anthony G.O. Yeh and Fiona F. Yang

44 **Locating Right to the City in the Global South**
Tony Roshan Samara, Shenjing He and Guo Chen

45 **Spatial-Economic Metamorphosis of a Nebula City**
Schiphol and the Schiphol region during the 20th century
Abderrahman El Makhloufi

46 **Learning Transnational Learning**
Edited by Åge Mariussen and Seija Virkkala

47 **Cultural Production in and Beyond the Recording Studio**
Allan Watson

48 **Global Perspectives on Gender and Space**
Edited by Ann M. Oberhauser and Ibipo Johnston-Anumonwo

49 **Fieldwork in the Global South**
Ethical challenges and dilemmas
Edited by Jenny Lunn

50 **Intergenerational Space**
Edited by Robert Vanderbeck and Nancy Worth

51 **Performativity, Politics, and the Production of Social Space**
Edited by Michael R. Glass and Reuben Rose-Redwood

52 **Knowledge and the City**
Concepts, applications and trends of knowledge-based urban development
Francisco Javier Carrillo, Tan Yigitcanlar, Blanca García and Antti Lönnqvist

53 **Migration, Risk and Uncertainty**
Allan M. Williams and Vladimír Baláž

54 **Transport, Mobility, and the Production of Urban Space**
Edited by Julie Cidell and David Prytherch

55 **International Perspectives on Age-Friendly Cities**
Edited by Francis G. Caro and Kelly G. Fitzgerald

International Perspectives on Age-Friendly Cities

Edited by Francis G. Caro and
Kelly G. Fitzgerald

Routledge
Taylor & Francis Group
LONDON AND NEW YORK

First published 2016 by Routledge

2 Park Square, Milton Park, Abingdon, Oxfordshire OX14 4RN
711 Third Avenue, New York, NY 10017

Routledge is an imprint of the Taylor & Francis Group, an informa business

First issued in paperback 2018

Library of Congress Cataloging-in-Publication Data
CIP data has been applied for

ISBN: 978-0-415-72071-7 (hbk)
ISBN: 978-1-138-54636-3 (pbk)

Typeset in Sabon
by Apex CoVantage

Contents

Box

Figures

Figures

Tables

Foreword

International Perspectives on Age-Friendly Cities

John R. Beard and Lisa Warth

The urban environments in which people live and grow older play significant roles in shaping their experiences of aging and of being an older person in the city. The natural, built, and social environments, including provided services, determine whether older people with given levels of function can continue to undertake the activities important to them. Barriers in the built environment can reduce older persons' mobility and constrain their opportunities to participate in social, economic, cultural, and spiritual lives. This in turn increases their risk of social isolation and loneliness. Physical barriers also increase the risk of falls and injury; they have negative impacts on health in other ways, too, for example, by discouraging physical exercise. Barriers in the social environment, such as ageist attitudes or services not adapted to the changing capacities, needs, and preferences of older people, may prevent them from getting the support and care they need and from actively contributing to and shaping community life.

Age-friendly cities identify and address such barriers and create opportunities for active and healthy aging. Doing so requires both political commitment and the active involvement of older people, caregivers, and service providers across all sectors in identifying needs and priority areas for action, implementing change, and evaluating progress over time. Cities and communities across the world that are participating in the WHO Global Network of Age-Friendly Cities and Communities have committed politically to making their communities more age-friendly, actively involving older people in the process. The WHO Age-Friendly Cities framework, which spans the built and social environments, highlights the need for collaborative thinking and coordination across sectors to ensure that improvements are coherent, equitable, sustainable, and centered on older people. Opportunities for participation not only need to be created, but also must be relevant, known about, and accessible. Over 250 cities and communities in 27 countries are currently applying and testing this approach to creating more age-friendly environments.

Enabling urban environments benefits people as they age by better enabling them to do the things that are meaningful for them. They benefit the wider community by enhancing opportunities for older people's active involvement and contribution to community life. Measures that benefit older people often benefit all generations. Older people are a heterogeneous group with diverse levels of capacity, diverse lifestyles, and diverse needs. Responding to these will serve urban dwellers at various stages of their lives. All citizens benefit from barrier-free design, safe public transport and public places, opportunities to rest in public areas, and respectful attitudes.

Creating age-friendly cities, therefore, implies creating enabling cities that encourage healthy and active aging for all. With the majority of the world's population living and aging in cities, there is a growing awareness and need to prepare the world's cities for the changing needs of aging populations. This book is therefore very timely, providing international perspectives on how cities have pursued the challenges of creating enabling environments. Case studies from Europe, Asia, and North America will provide the reader with different models and approaches to creating local structures and services adapted to the needs and preferences of older people. In the developing field of age-friendly city research and evaluation, this book makes an important contribution by presenting some of the insights gained and lessons learned so far.

Acknowledgments

Robert Geary made an enormous contribution to the project by managing the editorial process. He sought out the reviewers whom Kelly Fitzgerald and I nominated, facilitated communication between reviewers and authors, kept track of the status of manuscripts, and assisted authors with copy editing. By helping authors to tighten their manuscripts, he enabled us to expand the number of chapters that we were able to include in the book.

Since 2000, Robert Geary has been the managing editor of the *Journal of Aging & Social Policy*. In this capacity, he managed the editorial process for the special issue of the Journal that was the starting point for this book. In fact, he has played a similar role with three other edited books that began as special issues of the Journal. He has also assisted other members of the JASP editorial board, Edward Miller and Christian Weller, with special issues of the Journal that were published later as books. The current volume is the first in which articles from the special issue of the journal have been combined with new material in the book version. Mr. Geary also guided the editorial process for ten other special issues of the journal that were not published in book form. In this project and others, Mr. Geary has been particularly helpful in refining manuscripts for authors whose first language is something other than English.

Since Mr. Geary has announced his plans to retire from his position in the Gerontology Institute at the University of Massachusetts Boston at the end of 2015, this book will be his final large project. I welcome this opportunity to acknowledge his substantial contributions and thank him on behalf of the many authors who have benefitted from his assistance.

Francis G. Caro

Acknowledgments

Introduction

International Perspectives on Age-Friendly Cities

Kelly G. Fitzgerald and Francis G. Caro

Efforts to make cities and communities more age-friendly have gained significant momentum in recent years. Population aging and increased urbanization have challenged governments and other civic organizations to consider how best to develop a community that supports the full participation of residents regardless of age.

The growing interest in the age-friendly movement is the impetus for this book. Our work on age-friendly communities began with a special issue of the *Journal of Aging & Social Policy*. Because of the very strong response to our call for papers for the journal, we proposed an edited book on age-friendly cities and communities. This introduction provides a general overview to current age-friendly city elements, features, and initiatives that have influenced the age-friendly movement, and it briefly discusses opportunities and challenges that communities and their stakeholders might encounter in their age-friendly efforts.

The chapters in this book provide an international perspective highlighting different aspects of age-friendly communities. The book is divided into two sections: case studies and special topics. Some of the chapters were originally published in the special issue of the *Journal of Aging & Social Policy* (2014) mentioned above. A few of them have been slightly revised and/or updated from original publication. Other chapters, including most of the case studies, are new and provide a closer look at how cities have embraced age-friendly work. The special topics chapters that follow the case studies provide discussions of age-friendly program models, initiatives, research tools, and public policy issues. The sequence of the chapters is largely arbitrary; however, we grouped the case study chapters based on geographic location, beginning in Europe and Asia and crossing over to North America. Grouping them by location allows the reader to create some comparison of age-friendly work based on a region. This can be especially important when considering cultural, political, and economic conditions, all of which can influence age-friendly work in a region. In addition, some of the case studies provide cross-cutting themes highlighted at the beginning of that section.

Age-friendly initiatives have been introduced in places large and small. For that reason, the literature uses both the terms "city" and "community." In some cases, age-friendly initiatives are concerned with large municipalities as a whole. In other cases, they are concerned with neighborhoods or districts within large cities or in rural communities. Although the literature is quite broad when defining an age-friendly initiative, for the purpose of our review, and similar to Scharlach's (2012) literature review of U.S. initiatives, we will not include age-segregated housing or specialized programs. In discussing communities, we limit ourselves to geographic communities. (We exclude communities of interest or virtual communities.)

The chapters in this book focus primarily on how cities and communities in developed nations have and can become more age-friendly. Although several themes discussed in this issue can be adapted for developing nations, it is likely that barriers such as politics and economics severely limit how much a community can successfully accomplish. However, communities need not have substantial funds to institute age-friendly initiatives that can result in significant positive changes.

FEATURES OF AN AGE-FRIENDLY CITY OR COMMUNITY

An age-friendly city offers a supportive environment that enables residents to grow older actively within their families, neighborhoods, and civil society and offers extensive opportunities for their participation in the community. Alley, Liebig, Pynoos, Banerjee, and Choi (2007) define an age-friendly community as "a place where older people are actively involved, valued, and supported with infrastructure and services that effectively accommodate their needs" (p. 4). Age-friendly communities provide at least two pertinent forms of participation for older residents. In the more basic form, older people are simply active in the life of the community through interaction with other people and using community resources. This may take the form of driving, shopping locally, using parks and libraries, or attending church. Older people can also participate in political and related organizations within the community. This means consultation on the part of public officials with older people on matters that affect the development of different aspects of an age-friendly community.

An age-friendly community combines features—both environmental and social—to create living spaces that offer people rich opportunities for active aging. These aspects should be complementary and mutually reinforcing to allow older people to continue living and participating in the community.

Importance of the Age-Friendly-City Theme

Active aging is defined by the World Health Organization (WHO) as "the process of optimizing opportunities for health, participation, and security in order to enhance quality of life as people age. It applies to both individuals and population groups" (WHO, 2013b). When "active aging" is embraced, life expectancy can be extended, and individuals have opportunities to experience social, physical, and mental well-being while remaining connected with their communities (WHO, 2013a). Intergenerational and multigenerational interactions can also highly contribute to active aging.

Building on its active aging framework, the WHO began focusing on "age-friendly cities" in 2006 by conducting a focus group research project that addressed the concept of age-friendly cities and city living by older adults (Plouffe & Kalache, 2010; WHO, 2007). The result of the project was the development of a guide for cities to use as a tool to help identify age-friendly city features (WHO, 2007). Publication of the guide led to the development of the current WHO Global Network of Age-Friendly Cities and Communities. Although it is not the first or only organization focused on age-friendly developments, through these various recent efforts the WHO has become a particularly important resource for defining and describing age-friendly cities and communities.

To create a positive living environment for older people, the natural and built environments, social systems, participation, health, and safety are key elements to be considered. The intersection between quality of life and community design is important when planning for an aging population. Research points toward the influence of neighborhood design and the physical environment on quality of life (Lui, Everingham, Warburton, Cuthill, & Bartlett, 2009). A person's activities are influenced not only by the physical environment, but also by the housing, health, and social care services that are put in place to manage the needs of older adults. Personal empowerment and high functionality reduce dependency of older adults on already often-strained systems (WHO, 2007). When an older person experiences active aging, health and life expectancy are expected to improve, potentially reducing pressure on services such as health care. This is especially important in areas such as rural and remote communities in which the availability of services can be limited.

Collaboration among public health professionals, architects, urban design professionals, and others has increased on age-friendly community initiatives over the past decade. Although the focus is often more on the "built environment," and "aging" is only occasionally mentioned, there is overlap between some of the major age-friendly themes and initiatives that promote livable cities (Jackson, Dannenberg, & Frumkin, 2013). This overlap supports the argument that an age-friendly community is potentially "friendly" for all and not only older people.

An example of how collaborations are important for age-friendly projects can be found in Garon and colleagues' chapter about a medium-sized city in Quebec, Canada. It describes how the key to a successful initiative is formal collaboration among stakeholders (public and private): municipal government, political leaders, and representatives of groups representing seniors. Through a steering committee, the municipality devised an action plan that was integrated into the municipal service agenda. The steering committee consistently sought consensus among stakeholders. The pluralistic, participatory, and collaborative approach led to useful projects in a number of sectors. Some of these initiatives involve collaboration among units not accustomed to working together. A program to promote physical exercise for seniors that involved collaboration among 22 public and private organizations is offered as an example.

In their chapter on Portland, Oregon, DeLaTorre and Neal describe a collaboration among university gerontologists, city officials, and local senior advocates that led to the inclusion of senior concerns in the city's comprehensive plan. The formal planning process laid the groundwork for initiatives that will be helpful in the future. The chapter discusses how broad, cross-sector collaboration that has a sustained infrastructure is important. Sustainability and equality are also mentioned as important values that guide planning.

Keyes and colleagues' chapter provides an overview of the Lifelong Communities model that the Atlanta Regional Commission (ARC) introduced to assist communities in accommodating an aging population. Through collaboration among diverse professionals, the ARC is addressing housing, transportation, and health care issues. Application of the framework has led to a variety of local initiatives, including new community gardens and a farmers' market, better sidewalks and safer intersections, and zoning changes to facilitate development of new senior housing.

KEY ELEMENTS OF AN AGE-FRIENDLY CITY OR COMMUNITY

Developing an age-friendly environment requires the presence of certain preconditions and progress in achieving core and secondary age-friendly features. Preconditions are elements that must be in place before it is realistic to pursue an age-friendly agenda. Age-friendly initiatives are expected to be multidimensional. The most important dimensions are classified as core features. Potential agenda items, desirable but not essential, are classified as secondary features.

Our premise is that some basic community conditions must be in place if cities and communities are to pursue meaningful age-friendly initiatives. The necessary minimum level of favorable community conditions can be achieved through a combination of well-planned outdoor spaces, buildings, housing, transportation, sound economic activity, law-abiding behavior of residents, and well-operated municipal services. The premise is also that political institutions are in place to identify public needs, generate tax revenues, and deliver services to meet needs. In some countries, a network of established nonprofit organizations will be in place to address community needs.

Precondition Elements

Population Density

Large and growing concentrations of older people provide a rationale for making cities and communities age-friendly. The size and growth rate of the older population drive concerns about the importance of accommodating the older population. The number of people aged 60 and older is expected to double from 2006 to 2050 (Plouffe & Kalache, 2010). Over half of the world's older population lived in urban areas in 2005 (United Nations, 2010). The percentage of older people in urban areas was higher for developed nations compared to developing nations (United Nations, 2010).

Interest in older people in urban areas is increasingly becoming important in countries in which there have been large rural-to-urban migrations. Migrants from the rural areas have aged and continue to live in the cities to which they moved as young adults. Urban areas also include those who migrated from rural areas in old age to join adult children living there who migrated before them. Buffel, Phillipson, and Scharf (2012) argue that "developing new forms of urban citizenship will be especially important given the growth of migrant populations within urban areas" (Buffel et al., 2012, p. 607). Therefore, age-friendly initiatives must consider the special needs and circumstances of the aging migrant population especially because older migrants are at risk of being disadvantaged by precarious living conditions in urban areas (Buffel et al., 2012).

Climate and Weather

Climate can affect the types of activities in which older adults engage. For example, extreme climatic temperatures can limit an older person's ability to engage in outdoor activities. Communities that adapt structures and services to the climate offer older adults opportunities to continue to be active in the community. For example, Montreal, Canada, has an "underground city" with connections to transport, apartments, shopping, and other facilities, which protects residents from harsh winter conditions. Extreme weather events and disasters like hurricanes or floods affect how older people prepare for and community services respond to their needs. For example, emergency shelters designed to shelter older people might include access to special health care provisions or easily accessible beds. During the Hurricane Katrina response, the SWiFT Tool was implemented and found useful for triaging older adults in a large emergency shelter and potentially for pre-disaster planning (Mixson, 2007).

Topographical Features

Communities located on steep hills like the favelas in Rio de Janeiro, Brazil, or in low-lying areas like bayou communities in Louisiana in the United

States have unique topographical features that can present challenges to developing an age-friendly city or community. Topographical features are often overcome by local transportation or housing structures; however, creating a walking environment for an older person, for example, might prove difficult. An escalator that ascends 1,260 feet (approximately 384 meters), connecting the city center to a hillside residential area in Medellin, Columbia, is a recent example of how one community improved access for all residents, reducing walking time and offering accessibility to an area that might otherwise be difficult for older people to reach.

Social and Civic Organization

Contemporary age-friendly city initiatives build on a premise that sound social organizations and political institutions are already in place. Social organizations and political institutions are critical links between community conditions and the morale of residents (Ross, 2011). Communities with high concentrations of negative conditions illustrate the important contributions of social organizations and political institutions. Negative conditions include crime, deteriorated buildings, uncollected trash, and widespread graffiti.

Negative conditions in communities can be ameliorated through both informal and formal organizations. Civic action can also be pursued to achieve positive community conditions that enrich the lives of residents. Civic action that creates positive conditions can build confidence and trust among residents and may even encourage more "orderly" behavior (Ross & Jang, 2000).

Health and Social Services

Some age-friendly city initiatives address the needs of older residents for effective health and social services. Advanced societies are characterized by a rich array of specialized health and social services that have come to be recognized as essential for health and general well-being. Access to needed services and negotiation of complex health and social service systems can be challenging for people of all ages. The challenges are often particularly great for older people because of multiple chronic health conditions, disabilities, and modest financial resources. The potential barriers to use of health and social services include cost, information, and a combination of distance and transportation. Some age-friendly initiatives are addressing directly the challenges experienced by older residents by offering improved information about services, seeking to reduce cost as a barrier, and providing improved transportation. Others are seeking to improve the organization of services by seeking improved cooperation between public and private organizations involved in health and social service delivery. Some efforts involve better working relationships between municipal and regional governments, some between units of municipal governments.

Core Features

Housing

To offer people the possibility to age in the community, housing structures in a community should be varied (single family, apartments, assisted living, etc.) to meet the needs of not only different incomes but also those with various disabilities (MetLife Mature Market Institute, 2013). New housing units should incorporate universal design features. Examples include a no-step entry into the home, living space on one level, lever doorknobs, light switches located low enough for a wheelchair user, and walk-in showers. Similarly, communities can proactively approach future developments by requiring new housing to be developed with visitability and accessibility features (MetLife Mature Market Institute, 2013). Singapore, as discussed in the chapter by Chong and colleagues, has attempted to incorporate universal design in its densely populated area. Chong and colleagues focus on improving housing options with creative features, given that land is scarce in Singapore. Similar to universal design features, a home that has "visitability" is one that has three core features: a zero-step entrance, wider doorways, and at least a half bath on the main floor (Maisel, Smith, & Steinfeld, 2008). Visitability principles promote accessibility and eliminate barriers (Pynoos, Caraviello, & Cicero, 2009). Universal design principles and visitability principles are important for those who live in the home or who visit and have mobility difficulties.

Land-use zoning and property taxes can influence the availability and occupancy of various types of housing. Such codes specify the types of building structures and land development allowed in an area. These zoning codes influence the potential or current development of housing structures, ranging from single-family homes to facilities that provide group accommodations for those with extensive health care and support service needs. High property taxes can limit housing options for lower-income older adults by making some housing unaffordable for seniors. Age-friendly city initiatives may include strategies to reduce property taxes for low-income seniors. Zoning and tax codes can be changed by community leaders if the need presents itself; however, this can be an extended process. Significant property tax breaks for low-income seniors are also difficult to achieve because of the reliance of municipalities on property tax revenue and the reluctance of other property tax payers to subsidize low-income seniors. Nevertheless, there are creative strategies that communities can use that provide targeted tax relief without adding a significant cost burden for other residents. Keyes and colleagues' chapter describes how one community addressed zoning issues to improve housing availability.

The effort to introduce age-friendly initiatives in Philadelphia builds on a series of previous local efforts to address the needs of older people. Philadelphia is notable for giving particular attention to poor neighborhoods. Reflecting a special concern for vulnerable seniors and high housing costs, Glicksman and Ring's chapter discusses how Philadelphia has done significant work on zoning reform to permit accessible dwelling units.

Mobility

To remain active in a community, older people rely on various forms of transportation other than private automobiles. These include public transport, senior transportation services, and walkable neighborhoods (MetLife Mature Market Institute, 2013). In rural areas, transportation can be limited, for example, potentially isolating older people once they stop driving their own vehicles. Communities can help people drive safely and longer by making modifications to street signs, lighting, and other visibility features (MetLife Mature Market Institute, 2013).

Walking is a form of transportation regularly used by those who live in densely populated areas and can be particularly important for those who do not drive. Walking can also be a central strategy in the pursuit of a healthy lifestyle. In the United States, "complete streets" policies are one way in which some communities have implemented changes toward improving the walkability of an area (Lynott et al., 2009). The complete streets initiative encourages transportation planners to consider roadways from the perspective of all users, including bicyclists, public transportation vehicles, and pedestrians of all ages and abilities (National Complete Streets Coalition, 2013). Complete streets promote pedestrian safety, healthy lifestyles, and clean air and reduces the need for fossil fuels. In communities in which walking is already a highly used form of transportation, sidewalks and crosswalks may need to be constructed, may require a safety assessment, or may need to be repaired to ensure that they are accessible for older or disabled users. Communities may also need to regulate the use of sidewalks to assure that they are safe for pedestrians, such as by finding safe options for bicyclists on streets so they do not operate bicycles on sidewalks. Using Louisiana as an example in their chapter, Fields and Tudor provide an analysis of opportunities and barriers to changing transportation policies to become more age-friendly by implementing a Complete Streets policy. AARP played a major role in persuading the Louisiana Legislature to adopt a complete streets policy that has encouraged communities throughout the state to undertake specific complete streets initiatives.

Outdoor Spaces and Buildings

Independence, mobility, quality of life, and safety are affected by characteristics of the built environment (WHO, 2007). Outdoor spaces and buildings are two major features of a community that influence the ability of a person to age in place, yet they require significant forethought. The involvement of older people in the planning process can be particularly helpful to ensure that spaces and buildings meet the potential needs of older users. They can help to identify challenges and barriers for older people in current structures, contribute to the implementation or monitoring of age-friendly changes, and make recommendations for future spaces. Knowledge of aging and older people's needs among city planners, architects, property developers, and related others is also crucial to ensure that renovation and development of new spaces meet the needs of the aging population.

Participation of Older People

Older people participate in formal and informal activities as ways to maintain or establish relationships with colleagues, friends, and family; feel respect and esteem; and exercise competence (WHO, 2007). Access, variety, affordability, awareness, encouragement to participate, integration of generations, and culturally diverse communities are key factors to how much an older person participates in the community (WHO, 2007). Employment and volunteer opportunities are important to older people and are economically and personally beneficial (WHO, 2007).

Participation of older people in the development of age-friendly initiatives can take a number of forms. At minimum, information can be sought from older people through community hearings, focus group interviews, or formal survey research. Older people may be involved in gathering information. For example, they may be engaged as interviewers for assessments of needs of older residents. The age-friendly initiative in Portland, Oregon, is an example of a program that utilizes a university-city-community partnership (Neal, DeLaTorre, & Carder, 2014). Through activities including participatory research, this partnership offered older people opportunities to participate in the city's needs assessment to help identify ways in which Portland could become more age-friendly (Neal et al., 2014). Older people may also be included on boards and advisory committees that provide policy direction to age-friendly initiatives or take the lead in launching and directing age-friendly initiatives. Organizations representing older people may propose age-friendly initiatives and negotiate strong policy roles for them with other local organizations in the governance of age-friendly initiatives. Garon and colleagues (2014) provide examples of how older people participated at various stages of age-friendly projects in two communities in Quebec, Canada. Buffel et al. (2012) provide additional examples of how older people can participate in developing an age-friendly city or community. In some places, political institutions that address the needs of older people at local levels are well established. In Massachusetts in the United States, for example, the legislature in 1954 enacted legislation calling for the creation of municipal councils on aging (Massachusetts Commission on Aging, 1967). Councils on aging have a broad mandate to identify the needs of older people, to elicit participation of all residents in addressing their needs, and to promote and support programs that benefit older people in the community. Many of those appointed to serve on these councils are older people.

Volunteerism, however, is not always so straightforward. Tuymazy (Russian Federation) had not participated in previous, externally conceived initiatives focused on the needs of older people (Minnigaleeva's chapter in this book). The case study chapter on Tuymazy and Bashkortostan demonstrates the importance of the historical context. Tuymazy remains in transition from the Soviet era. Residents continue to have the expectation that their needs will be met by government. Because volunteering was mandatory in the Soviet era, residents tend now to have an aversion to volunteering. Tuymazy followed closely the organizing methodology recommended by the WHO age-friendly initiative. Credibility and respect for the WHO gave them strength to develop their age-friendly work. (This is an essential piece for many developing nations that want to pursue this work.) Following a slow and deliberate process to organize a nongovernmental organization (NGO) that introduced age-friendly programs, citizens and public officials were eventually receptive. The initiative introduced some new services that could be operated with modest funding. In a context in which little had been available previously, the new services were very well received by residents.

Shannon and O'Connor's chapter discusses how attention to the needs of the aging population is relatively recent in Ireland. A national positive-aging initiative set the stage for the Aging Well initiative. This initiative is active in 17 cities and regions, with others likely to join soon. The program starts with a consultation process, which includes focus groups with a range of people, with a view to learn priorities of older people. The outcome is the basis for the strategy. A survey is also used in which older people perform as researchers.

Secondary Features

Age-Friendly Businesses

Many age-friendly programs have begun to adopt age-friendly business strategies. Often there is no or very little cost for businesses to make changes to become more age-friendly, yet the reward for doing so can be a potential increase in the business's value in the community and its revenue. New York City, counties in Ireland, and Bowling Green, Kentucky, all differently sized communities, have each implemented age-friendly business programs that either award or recognize businesses that use age-friendly practices. Examples of age-friendly features might include assisted devices to open doors, access to restrooms with easily reachable toilet paper dispensers, reduced background music, increased font size on menus or advertisements, and clear oral communication used by staff. Some facts that support the benefit of implementing age-friendly business practices are that older consumers will double with the aging population; they tend to outspend younger consumers; and they tend to be more loyal customers (Age Friendly Counties Ireland, 2013). An example of a New York City initiative is the Age-Friendly Local Business Initiative, with more than 1,000 participating businesses. The initiative is an educational outreach campaign that includes self-assessment materials, a shop window decal, and inclusion on a list shared with older consumers so that businesses can increase their number of older customers with little or no cost (Age-Friendly NYC, 2013). The city of Bowling Green gave its first Age-Friendly Awareness Business Award in August 2013 to a nominated business that had taken recognized action towards becoming more age-friendly (WKU Center for Gerontology, 2013). Quickly becoming an annual tradition, in 2014 Bowling Green awarded its second Age-Friendly Awareness Award.

Measuring Age-Friendliness

For communities, "age-friendly" is more usefully considered a quality to be sought than an end state that has been achieved. Age-friendliness is a multidimensional construct. The dimensions (health, housing, participation, and so on) have no practical ceilings. No standards have been established for classifying the degree to which communities have achieved age-friendliness on either component dimensions or a summary measure. Because of inevitable uncertainties about the bases for establishing standards and the extreme variability among communities, organizers of age-friendly initiatives are unlikely to propose standards. Further, organizers are unlikely to certify communities as "age-friendly" for an end state that they have achieved.

Nevertheless, the dimensions that contribute to age-friendliness can be measured. Communities are encouraged to measure conditions and to monitor changes. Communities may assess themselves on the degree to which they have become more (or less) age-friendly. For example, the WHO is currently developing a measurement tool that offers a core set of indicators for communities to use. We can also anticipate some comparisons across municipalities that rate them according to the degree to which they have become more (or less) age-friendly.

EXAMPLES OF CONTEMPORARY INITIATIVES TO ENCOURAGE COMMUNITIES TO STRENGTHEN AGE-FRIENDLY FEATURES

Globally, there are various approaches that organizations use to promote age-friendly initiatives. The following are a few highlighted examples of initiatives that encourage the development of age-friendly cities and communities. For further information regarding initiatives, see Scharlach (2012), Grantmakers in Aging (2013), and Lui et al. (2009).

The WHO Global Network of Age-Friendly Cities and Communities

The WHO Global Network (abbreviated) began in 2010 as a network of cities around the world using an active aging framework focused around eight domains to address environmental and social factors that contribute to an aging community (WHO, 2012). As of December 2014, there were 11 affiliated programs and more than 250 cities across 27 countries that have joined the network. covering an astonishing over 88 million people.

Each of the member cities is required to complete a five-year cycle focusing on essential features or domains that contribute to structures and services that are accessible and inclusive of older people (WHO, 2013a). Affiliate programs support the network in varying ways and include opportunities for cities located in the affiliated programs' regions to join the WHO Global Network either directly or indirectly through a national government-led program. The WHO Global Network provides cities the opportunity to network with each other over an Internet platform called Age-Friendly World (www.agefriendlyworld.org*)* and will soon have a core set of indicators available for cities and communities to use as they develop their individual age-friendly programs. Several chapters in this book discuss participation in the WHO Global Network and the application or implementation of the WHO domains.

In contrast to other manuscripts in this collection, Lopes and Lemos offer a methodological contribution that tests the WHO's age-friendly cities and communities tools. To assist local planners, the WHO developed a checklist organized in eight domains. Using the checklist as the basis for a survey by older people of communities in Portugal, Lopes and Lemos show that the domains in the guide are multidimensional. The findings imply that municipal planners not only need to integrate activities across the major domains, but also need to be sensitive to complexity within domains.

European Commission

The European Commission recently created the Action Group D4 on Age-Friendly Environments as part of the European Innovation Partnership on Active and Healthy Aging, which has approximately 70 commitments from partner organizations. An example of one accomplishment includes a report published in 2013 that stated that age-friendly work was high on the agenda for municipalities, lack of funding was an obstacle, and priorities included sharing good practices and information on funding (European Commission, 2014). In cooperation with other groups, a thematic network called AFE-INNOVNET (http://afeinnovnet.eu/) and a proposal for a "European Union covenant on demographic change" have also been developed, along with a few other accomplishments.

Since approximately 2013, Age-Friendly Environments in Europe (AFEE), a joint project between the European Commission Department of Employment, Social Affairs, and Inclusion and the WHO Regional Office for Europe, began in order to create tools for authorities to help them commit to age-friendly work and to measure their progress (World Health Organization Regional Office for Europe, 2014).

In the United States, a number of distinct initiatives have emerged to promote age-friendly themes. One is sponsored by the Environmental Protection Agency, an agency of the federal government. Others are sponsored by private, nonprofit organizations: AARP, the Visiting Nurse Service of New York, and the National Association of Area Agencies on Aging.

AARP Livable Communities

AARP uses the livable communities approach, in which it "seeks to improve the quality of life for older adults by promoting the development of safe, accessible, and vibrant environments often called livable communities. Livable communities policies address issues such as land use, housing, and transportation, which are vital to developing communities that facilitate aging in place" (AARP, 2013). The purpose of this program is to identify strategies and guidelines for developing environments that embrace successful aging (Scharlach, 2009). It offers publications, toolkits, videos, and group discussions around various topics related to livable communities.

The Visiting Nurse Service of New York's AdvantAge Initiative

The AdvantAge Initiative began as a community-building effort to create elder friendly or "AdvantAged" communities to meet the needs of older people (AdvantAge Initiative, 2013). This initiative is organized by the Center for Home Care Policy and Research of the Visiting Nurse Service of New York City. The major activity for this initiative was a survey of how older adults fare in their communities and focused on how older adults think about their communities. The survey used four domains focused on housing and security; physical and mental health; independence for the frail, disabled, and homebound; and opportunities for social and civic engagement. The initiative has been implemented in over 50 communities (Oberlink, 2014) across the United States and includes five stages of activities. The initiative has developed 33 indicators, or pieces of information, that are tracked over time, and benchmarks (which show change in an indicator over time) to help communities evaluate their age-friendliness (AdvantAge Initiative, 2013).

U.S. Environmental Protection Agency's Building Healthy Communities for Active Aging (BHCAA) Award Program

Following the launch of the 2002 Aging Initiative and other related activities, the United States Environmental Protection Agency (EPA) developed the BHCAA award program to recognize communities in the United States that have combined smart growth and active aging concepts. The principles of smart growth focus on how the built environment affects the ability of people to live and access resources in a community (U.S. EPA, 2012a).

According to the EPA, the impact that developments have on the environment is crucial for air, water, brownfields, and open-space preservation. Related more specifically to older adults is the walkability of a community, which might reduce the need for some transportation while allowing improved accessibility, and housing options that meet the varying needs of aging individuals that allow them to age in place. A community that utilizes smart growth aims to be both visually vibrant and utilitarian and to limit the impact of community development on the environment (U.S. EPA, 2012b).

The EPA presents awards to communities that use an inclusive approach at varying levels of governmental jurisdictions. Sykes and Robinson's chapter in this book describes the EPA's public education strategy to disseminate ideas about supporting an older population by awarding cities that have implemented initiatives in such areas as housing, management of patients with chronic diseases, and promotion of active aging. Since 2007, 21 communities in 15 states have won awards (U.S. EPA, 2011).

Glicksman and Ring's chapter provides an example of how Philadelphia, which has two age-friendly efforts, has drawn upon both the EPA and WHO models. Age-Friendly Philadelphia emerged from concerns about increased urban poverty and interest in neighborhood revitalization. The chapter discusses attempts to link the two programs together. Both initiatives have placed a good deal of emphasis on research. Both initiatives have also benefitted from a good deal of professional direction. They have tried to increase the network of traditional and nontraditional aging organizations.

National Association of Area Agencies on Aging's Livable Communities Initiative

The National Association of Area Agencies on Aging (n4a) focuses on livable communities and has joined with Partners for Livable Communities (Partners) in an Aging in Place Initiative, which is funded by the MetLife Foundation (n4a, 2013). N4a offers technical assistance, such as a guide for local leaders that provides tools to assist in collaborations, training, access to speakers, workshop development, and focus groups to help organizations develop livable communities. For example, 12 workshops enriched by small grants were made available from 2007 to 2009 by Partners and n4a. The grants were used by partnerships that focused on innovative action plans for aging in place in communities (Partners for Livable Communities, 2012).

Village Movement

The purpose of a village is to help people to age in their homes and to remain active in their communities. Villages are membership-driven grassroots organizations that coordinate access to affordable services that enable people to age in their communities (Village to Village Network, 2013). The Village movement in some communities is more highly related to age-friendly initiatives than it is in other communities. The goal for the first village was to consolidate information about already available services so that residents could easily access them (McWhinney-Morse, 2009). There are now more than 120 villages across four countries, with approximately 100 others under development (Village to Village Network, 2014). A goal of these villages is that older adults can remain engaged in the community; formal and informal services are integrated, making them easier to access; civic engagement is promoted; and service delivery silos are removed (Greenfield, Scharlach, Graham, Davitt, & Lehning, 2012).

Scharlach and colleagues' chapter examines the potential link between the "Village movement" and the age-friendly city movement. Villages are private organizations developed largely by seniors to assist them and other members in living successfully in their homes and communities. In some instances, villages have broadened their scope to address the community-wide needs of municipalities with expanding numbers of older people.

OPPORTUNITIES AND CHALLENGES

Both opportunities and challenges will arise when cities and communities work toward becoming more age-friendly. Every city or community will have unique points to address; however, there are a few general opportunities and challenges that all cities and communities will have to consider to some extent.

Opportunities

Aging Population

The dramatic and rapid growth of the older population provides strength and urgency to the age-friendly movement. With 22% of the population expected to be older than 60 by 2050 (WHO, 2012), there will be an increased need for age-friendly structures and services. The changing expectations of forthcoming generations of older people to be active and to participate in the community will require a community environment that offers older people extensive and varied opportunities.

Automobile-oriented communities with widely dispersed housing and scattered commercial and cultural resources are challenged to achieve greater density. To retain their older population, communities need to encourage dense residential development near compact commercial and cultural centers close to public transportation. For some communities, this transformation will be very expensive. The shift to denser development will make communities more attractive to much of the younger population that is attracted to walking, bicycling, and using public transportation. Communities that offer older people extensive housing with universal design features in dense neighborhoods with extensive age-friendly features will need less age-segregated service-supported housing, such as nursing homes and assisted living.

Potential for Accomplishments on the Basis of Current Resources

To some extent, communities are likely to be able to become more age-friendly through thoughtful use of existing resources. Expensive new initiatives are not always necessary. In complex contemporary societies, for example, lack of information is often a barrier to accessing community resources. In these cases, access can be improved through more effective dissemination of information. The value of dissemination of information is exemplified by a guide to restrooms open to the public, developed by an age-friendly city program in Brookline, Massachusetts, United States. Need for frequent urination discourages some older people from leaving home. Information about the location of restrooms is not widely available. The Brookline program encourages participation among older residents by publishing a guide, available online, to the location of restrooms in the community that are open to the public.

In other cases, communities can become more age-friendly through better cooperation among municipal departments. In Brookline, collaboration between the town's transportation department and the town's Council on Aging has led to improvement in a program that provides better parking opportunities for those who care for older people in their homes. The Brookline Council on Aging disseminates information about the permit program that has been developed and offers help to caregivers in applying. No new financial resources were needed to operate the program. No doubt there are many ways in which age-friendly city initiatives may stimulate municipalities to take modest but useful steps to make the community more age-friendly. Individually, these steps probably have very modest impacts, but cumulatively they may be significant. The potential for taking useful but modest incremental steps through age-friendly initiatives does not preclude the need for examining the larger reform issues suggested by Ball and Lawler's chapter. Ball and Lawler argue that, to make our cities age-friendly, we must fundamentally rethink many of our assumptions about community life. An aging society requires major changes in housing stock, urban design, transportation systems, and health care systems. Some of our systems cannot be fixed through incremental changes. Instead, Ball and Lawler propose that outdated systems be dismantled and replaced by new systems that better reflect contemporary needs. The Ball and Lawler position is an important critique of the vast majority of age-friendly initiatives that involve modest refinements of established policies and programs.

"Age-Friendly" is Potentially All-inclusive

Why would a city focus on "old-age-friendly" and not consider another approach such as healthy cities, environment, child-friendly, or dementia-friendly? Many cities are pursuing other competing initiatives. However, all of the age-friendly initiatives currently available have some elements that benefit all ages and their needs and abilities. Some initiatives like "villages" have a stronger focus on services for older people; however, the general concept behind the program will benefit a larger proportion of the population, especially as the population ages.

Warner and Homsy argue in this book that age-friendly initiatives can be approached in a way that is age-inclusive. Warner and Homsy examine possibilities for communities to pursue initiatives designed simultaneously to address needs of seniors and children. They discuss, for example, redesign of schools in poor neighborhoods so that they become community centers serving people of all ages. They also discuss intergenerational programs that benefit both seniors and children. Further, they recommend participation of people old and young in the development of programs to benefit people of various ages.

Challenges

Economy and Government

Tight municipal budgets can be serious barriers to public support for age-friendly city initiatives. The public financing problem is particularly serious when economic conditions are generally weak because governments are challenged not only by declining revenues but also by increased demand for safety net services. In these circumstances, governments find it difficult to finance discretionary, future-oriented programs. In an era in which public resources for new initiatives are chronically scarce, age-friendly city initiatives may have to be launched with support from nongovernmental entities and conducted on a modest scale. They have to draw on public funds allocated for more basic purposes. They may also have to depend heavily on volunteers.

Menec and colleagues' chapter provides an example of a program with limited resources. As a province, Manitoba joined the WHO initiative. For municipalities, this was a top-down initiative, government-led. While many municipalities were cooperative in organizing age-friendly city initiatives, they were constrained by their modest fiscal resources. Some made modest changes to their infrastructure that were feasible without new money.

In developing nations or communities, a consistently poor economy or lack of support from government for aging programs might be close to impossible. In these countries, information about the needs of an aging population might not be available; therefore, support for an initiative such as an age-friendly community might be seen as beyond anything that could be attempted.

Politics and Policies

Some initiatives like the WHO Global Network require government participation. Support of top political officials is important because key age-friendly initiatives may be expensive or contentious. Public transportation systems, for example, typically need substantial and continuing public subsidies. Higher-density residential development, for example, is likely to require politically contentious zoning changes. Strong political leadership is needed to support major initiatives. Although sometimes challenging to get a government initially to sign on to age-friendly work, political players are constantly changing, which can result in changes in policy interests. With this comes a change in priorities and the potential for age-friendly work to cease. New York City, a member of the WHO Global Network, for example, attempted to address this issue by identifying priorities for the new mayor based on a consultation of experts. Smaller communities in which age-friendly work is grassroots-led may not be so severely affected by a change in political players and policies; however, there may be indirect effects that can affect the grassroots efforts.

Sustaining Age-Friendly Work

Sustainability of an age-friendly community program is a major challenge for any community. It is often easy to gain buy-in to the program because the value of the work is evident. However, it can be very difficult to continue momentum when politics, the economy, and society in general are constantly changing. As a community begins to develop its plan to improve its age-friendliness, it is essential to consider how to sustain the work beyond new leadership, especially if funding disappears, such as discussed in DeLaTorre and Neal's chapter. Although not every potential change can be foreseen, a plan should be in place to ensure that work continues. Otherwise, it is likely that when a roadblock is encountered, the momentum of work can be abruptly halted.

Challenges of Monitoring, Evaluation, Benchmarking, Recognition, and/or Comparability

Each of these activities is important to age-friendly communities if age-friendly work is ongoing and expected to succeed. It is essential for cities to monitor, benchmark, and evaluate the work that has been done to determine the extent to which the city has become more age-friendly. However, these types of activities often cost money to conduct and sometimes are left to the end to complete. If left to the end, funding can dry up or be redirected, leaving some of the most important activities incomplete. These types of activities also require competent people. For smaller communities, this can be especially challenging if there are no qualified individuals to complete the tasks or no financial resources to hire someone. Recognition and comparability are activities that can help support and provide legitimacy for city or community age-friendly activities; however, it can be difficult to offer or conduct them, especially if cities/communities in different countries are considered. What works in a city in India may not be useful or possible for a city in Canada. However, the chapter by Buffel and colleagues provides a comparison of two European cities, Brussels, Belgium, and Manchester, England, highlighting the different approaches taken in each city's age-friendly work. While there are some similarities in activities for both cities, such as having dedicated teams working on policy, older people involved in the process, and a focus on initiatives for social inclusion and participation, recognized differences were identified such as the promotion and perception of age-friendly city projects in the community, attention paid to specific groups, and a different focus on outdoor environments.

CONCLUSIONS

What the age-friendly city movement will accomplish in the long run will depend on many factors, including the energy, imagination, and resourcefulness of those involved. Since many age-friendly city initiatives are in their early stages, the possibilities for what they will accomplish are open-ended. Because of the complexities of contemporary communities, the potential for strengthening opportunities for participation of older people in various sectors of community life is enormous. The diversity of interests and skills of older people has implications for the range of opportunities that might be nurtured. In some communities, age-friendly city initiatives stimulate attention to other special populations such as children, adults with disabilities, and lesbian, gay, bisexual, and transgender (LGBT) older adults. To the extent that they may have limited access to their own transport, older people and children share an interest in good pedestrian features and public transportation. They also share interest in local parks but differ in the park features of interest. Adults with disabilities and those without disabilities have shared interests in safe sidewalks and specialized transportation. Age-friendly initiatives must consider the needs of all populations, including those traditionally marginalized, to help break down barriers associated with the label "special populations." For example, older LGBT persons might have an interest in the opportunity to access mainstream services that also respond to diverse identities. Services could include care facilities, meeting spaces, or support networks. In these communities, possibilities emerge for collaboration among those advocating on behalf of several population groups. A result may be initiatives addressed to a coalition of groups. The possibilities are endless. Constraints on financial resources will set some limits on what can be accomplished in some communities. On the other hand, modest projects launched by age-friendly city initiatives may lay the groundwork for larger, more generously financed initiatives funded by national governments.

REFERENCES

AARP. (2013). *Livable communities*. Available from http://www.aarp.org/research/ppi/liv-com.html

(The) AdvantAge Initiative. (2013). *What is the AdvantAge Initiative?* Available from http://www.vnsny.org/advantage/whatis.html

Age Friendly Counties Ireland. (2013). *Facts and figures*. Available from http://www.agefriendlycounties.com/age_friendly_business/facts_and_figures

Age-Friendly NYC. (2013). *Local business initiative*. Available from http://www.nyam.org/agefriendlynyc/initiatives/current/age-friendly-business.html

Alley, D., Liebig, P., Pynoos, J., Banerjee, T., & Choi, I.H. (2007). Creating elder-friendly communities: Preparations for an aging society. *Journal of Gerontological Social Work, 49*, 1–18.

Buffel, T., Phillipson, C., & Scharf, T. (2012). Ageing in urban environments: Developing 'age-friendly' cities. *Critical Social Policy, 32*, 597–617.

European Commission. (2014). *Action group D4: Age-friendly environments*. Available from http://ec.europa.eu/research/innovation-union/index_en.cfm?section=active-healthy-ageing&pg=action_group_d4

Garon, S., Paris, M., Beaulieu, M., Veil, A., & Laliberté, A. (2014). Collaborative partnership in age-friendly cities: Two case studies from Quebec, Canada. *Journal of Aging & Social Policy, 26*(1–2), 73–87.

Grantmakers in Aging. (2013). *Age-friendly communities: The movement to create great places to grow up and grow old in America*. Arlington, VA: Grantmakers in Aging.

Greenfield, E.A., Scharlach, A.E., Graham, C.L., Davitt, J.K., & Lehning, A.J. (2012). *A national overview of villages: Results from a 2012 organizational survey*. New Brunswick, NJ: Rutgers, The State University of New Jersey, School of Social Work.

Jackson, R.J., Dannenberg, A.L., & Frumkin, H. (2013). Health and the built environment: 10 years after. *American Journal of Public Health, 103*, 1542–1544.

Lui, C.-W., Everingham, J.-A., Warburton, J., Cuthill, M., & Bartlett, H. (2009). What makes a community age-friendly: A review of international literature. *Australasian Journal on Ageing, 28*, 116–121.

Lynott, J., Haase, J., Nelson, K., Taylor, A., McCann, B., & Stollof, E.R. (2009). *Planning Complete streets for an aging America*. Washington, DC: AARP Public Policy Institute.

Maisel, J.L., Smith, E., & Steinfeld, E. (2008). *Increasing home access: Designing for visitability*. Washington, DC: AARP Public Policy Institute.

Massachusetts Commission on Aging. (1967). *Design for action manual for councils on aging*. Available from http://www.mcoaonline.com/content/pdf/Enabling_Legislation.pdf

McWhinney-Morse, S. (2009). Beacon Hill Village. *Generations, 33*(2), 85–86.

MetLife Mature Market Institute. (2013). *Livable community indicators for sustainable aging in place*. New York, NY: Metropolitan Life Insurance Company.

Mixson, P. (2007). From Houston—The SWiFT tool and recommendations for best practices. *Victimization of the Elderly & Disabled, 9*(6), 81, 90–93.

n4a. (2013). *Livable communities*. Available from http://www.n4a.org/programs/livable-communities/

National Complete Streets Coalition. (2013). *National Complete Streets Coalition*. Available from http://www.smartgrowthamerica.org/complete-streets?key=31733746

Neal, M., DeLaTorre, A., & Carder, C. (2014). Age-friendly Portland: A university-city-community partnership. *Journal of Aging & Social Policy, 26*(1–2), 88–101.

Oberlink, M. (2014). Personal Communication.

Partners for Livable Communities. (2012). *Aging in place initiative–developing livable communities for all ages*. Available from http://www.livable.org/program-areas/livable-communities-for-all-ages-a-aging-in-place/the-aging-in-place-initiative

Plouffe, L., & Kalache, A. (2010). Towards global age-friendly cities: determining urban features that promote active ageing. *Journal of Urban Health: Bulletin of the New York Academy of Medicine, 87*(5), 733–739.

Pynoos, J., Caraviello, R., & Cicero, C. (2009). Lifelong housing: The anchor in aging-friendly communities. *Generations, 33*(2), 26–32.

Ross, C. E. (2011). Collective threat, trust, and the sense of personal control. *Journal of Health and Social Behavior, 52*, 287–296.

Ross, C. E., & Jang, S. J. (2000). Neighborhood disorder, fear, and mistrust: The buffering role of social ties with neighbors. *American Journal of Community Psychology, 28*, 401–420.

Scharlach, A. (2009). Frameworks for fostering aging-friendly community change. *Generations, 33*(2), 71–73.

Scharlach, A. (2012). Creating aging-friendly communities in the United States. *Ageing International, 37*, 25–38.

United Nations. (2010). *World population ageing 2009*. New York, NY: Department of Economic and Social Affairs, Population Division, United Nations.

U.S. Environmental Protection Agency. (2011). *Building Healthy Communities for Active Aging (BHCAA)*. Available from http://www.epa.gov/agingepa/bhc/

U.S. Environmental Protection Agency. (2012a). *Environmental benefits of smart growth*. Available from http://www.epa.gov/smartgrowth/topics/eb.htm

U.S. Environmental Protection Agency. (2012b). *About smart growth*. Available from http://www.epa.gov/smartgrowth/about_sg.htm

Village to Village Network. (2013). *Frequently asked questions*. Available from http://www.vtvnetwork.org/content.aspx?page_id=22&club_id=691012&module_id=70769&sl=164151088#What%20exactly%20is%20a%20Village

Village to Village Network. (2014). *About VtV network*. Available from http://www.vtvnetwork.org/content.aspx?page_id=22&club_id=691012&module_id=65139

WKU Center for Gerontology. (2013). *Age friendly cities project*. Available from http://www.wku.edu/aging/agefriendlycitiesindex.php

World Health Organization. (2007). *Global age-friendly cities: A guide*. Geneva, Switzerland: World Health Organization.

World Health Organization. (2012). *WHO age-friendly environments programme*. Available from http://www.who.int/ageing/projects/age_friendly_cities_programme/en/index.html

World Health Organization. (2013a). *What is "active ageing"?* Available from http://www.who.int/ageing/active_ageing/en/

World Health Organization. (2013b). *WHO global network of age-friendly cities and communities*. Available from http://www.who.int/ageing/projects/age_friendly_cities_network/en/index.html

World Health Organization Regional Office for Europe. (2014). *Age-friendly environments in Europe (AFEE)*. Available from http://www.euro.who.int/en/health-topics/Life-stages/healthy-ageing/activities/age-friendly-environments-in-europe-afee

Part I

Case Studies

Cross-Cutting Themes

Europe

1 Developing Age-Friendly Cities

Case Studies from Brussels and Manchester and Implications for Policy and Practice

Tine Buffel, Paul McGarry, Chris Phillipson, Liesbeth De Donder, Sarah Dury, Nico De Witte, An-Sofie Smetcoren, and Dominique Verté

INTRODUCTION

Developing age-friendly communities has become a significant dimension in debates in social policy (Buffel, Phillipson, & Scharf, 2012). A variety of factors has stimulated discussion around this topic, including, first, the impact of demographic change across the global north and south (Menec, Means, Keating, Parkhurst, & Eales, 2011); second, awareness of the impact of urban change on older people, notably in areas experiencing social and economic deprivation (Buffel, Phillipson, & Scharf, 2013); and third, debates about good or optimal places to age, as reflected in debates around lifetime homes and lifetime neighborhoods (Scharlach & Lehning, 2013).

The issue of developing age-friendly communities arose from a number of policy initiatives launched by the World Health Organization (WHO) during the 1990s and early 2000s. A central theme concerned the idea of active aging, originally developed during the United Nations' Year of Older People in 1999. Here, the notion of "active" was taken as referring to participation across a range of areas—social, cultural, spiritual, and economic—in addition to those associated with physical activity. Policies and programs directed at achieving active aging were considered to require a range of interventions, including actions at the level of the social and physical environments. This idea was taken further in 2006 when the WHO launched the Global Age-Friendly Cities project. (Please refer to the Introduction of this volume.)

Lui, Everingham, Warburton, Cuthill, and Bartlett (2009) suggest that a variety of approaches show how favorable environments for older citizens may be sustained, ranging from emphasis on physical infrastructure, on the one hand, to the quality of social relations that promote participation, on the other. There is also discussion in the literature about the governance processes necessary for defining and building age-friendly communities, with older people being seen to have key roles in contributing to their distinctive features (Lui et al., 2009; Menec et al., 2011). This implies the bottom-up participation and involvement of older citizens in voicing their concerns, as well as their participation in decision-making and planning. The emerging concept of "rights to the city" is also relevant when addressing the role of older people in developing age-friendly communities. This idea is linked with the work of Lefebvre (1991) and points to the need for developing the rights of citizens to participate in decision-making regarding the production of space and access to resources (e.g., in aspects such as public transportation and neighborhood regeneration (Purcell, 2003)). Such rights to the city may be particularly important for older people who become reliant on their immediate environments. Yet the evidence suggests that they are often the last to be engaged when it comes to decision-making processes within their neighborhoods (Buffel et al., 2012). The global age-friendly movement (WHO, 2009) has been highly influential in raising awareness of the potential of involving older people in shaping strategies for urban planning and community regeneration. However, this approach is a relatively new development and has yet to receive systematic evaluation. In particular, research is needed that identifies the various approaches to building age-friendly communities relevant to different cultural and sociopolitical contexts (Menec et al., 2011).

Given the above, this chapter aims to provide a comparison of the age-friendly approaches in two European cities, Brussels and Manchester, with a particular focus on initiatives that promote active aging in an urban context. Both cities are members of the WHO Global Network of Age-Friendly Cities and Communities (WHO Global Network) and have demonstrated their commitment to "a cycle of continually assessing and improving their 'age-friendliness' " (WHO, 2009). The opportunity to compare these cities arose from collaborative work between researchers from Brussels and Manchester investigating the impact of social deprivation in the respective cities (Buffel et al., 2013; Buffel & Phillipson, 2011). Three important benefits of a comparative study can be identified: first, it generates knowledge about the various ways in which the age-friendliness of a city can be assessed, drawing on contrasting urban environments. Second, a comparative approach is valuable in improving our understanding of success factors and barriers to implementing age-friendly programs across different sociopolitical, cultural, and policy contexts. Third, it may foster mutual learning between cities and inspire others striving to meet the needs of their older residents.

This chapter examines the demographic, social, and multicultural contexts of Brussels and Manchester; the ways in which both cities became members of the WHO Global Network; similarities and differences in the age-friendly approach and actions adopted by both cities; and opportunities and barriers to the implementation of age-friendly policies and initiatives.

The chapter concludes by discussing key elements and resources needed to develop age-friendly cities. Exploration of these areas will be used to develop a framework for examining the range of factors that can assist as well as impede implementation of the age-friendly approach.

THE DEMOGRAPHIC AND SOCIAL CONTEXTS OF BRUSSELS AND MANCHESTER

Brussels: Population and Social Dimensions

Brussels is both a region and a city. The Brussels-Capital Region is one of the three federated regions of Belgium, alongside Wallonia and the Flemish Region, and serves as the de facto capital of the European Union, hosting principal EU institutions as well as other international organizations such as NATO. It covers 161.4 square kilometers or half a percent of Belgium's territory, but the population is equivalent to more than 10% of the population (1,104,346 inhabitants recorded in 2010; Deboosere, 2010). The Region of Brussels-Capital, an enclave within the Flemish Region, is an official bilingual region. Both the Flemish and the francophone communities have jurisdiction in the Brussels-Capital Region. The Brussels-Capital Region comprises 19 autonomous municipalities, one of which is officially named the City of Brussels. It is the largest municipality, with 163,210 residents, of whom 16% (or 25,669 residents) are aged 60 and older. The City of Brussels is the official capital of Belgium (Deboosere, 2010).

Brussels is currently the fastest growing region of Belgium. Migration flows continue to be the largest influence on the region's composition, although a natural increase continues to be an important factor in population growth. Forty-six percent of the Brussels population is not of Belgian origin, with especially large populations of people of Moroccan, Italian, Spanish, Turkish, and Portuguese origins, alongside a rapid increase in immigration from Eastern European countries (e.g., Poland and Romania) and Sub-Saharan Africa. The Brussels-Capital Region has a relatively young age structure, with a constantly increasing birth rate and decline in the median age. The average age of residents in the City of Brussels (36.5 years) is even lower than in the Brussels Region (37.7 years), compared with 40.68 years for Belgium as a whole. Unlike the rest of the country, the number of residents aged 65 and older has decreased over the last 15 years in the Capital Region (–9%) (Brussels-Capital Health and Social Observatory, 2008). In contrast, the Brussels Region has proportionately more people aged 80 years and older. It is projected that the "aging of the aged" will further increase: by 2050, the number of older people aged 80 and older will account for 40% of the over-60s, compared with 27% in 2008 (De Spiegelaere, Closon, Deboosere, & Humblet, 2009).

A study of poverty in Brussels revealed a significantly higher percentage of older people reliant on the minimum income set for elders (7.5% compared with 4.8% in the rest of Belgium), an increase in the number of those living on very low incomes, and rising numbers of unemployed people aged 50 and older (Brussels-Capital Health and Social Observatory, 2008). Older people who live alone and who do not own their homes are especially vulnerable to problems arising from low incomes. The socioeconomic situation of those living in the inner city is especially precarious. At the same time, inequalities in the older population are intensifying, partly because of the growing numbers of older migrants and non-EU ethnic minorities, many of whom face higher risks of discrimination, poverty, and health problems. By 2020, it is expected that half of the older population in Brussels will be of foreign origin (Vanmechelen et al., 2012).

Manchester is a city and a metropolitan borough of Greater Manchester, England. Greater Manchester has a population of approximately 2.68 million and comprises 10 metropolitan districts, each of which has a major town center and outlying suburbs. Out of the 10 districts, the City of Manchester has the largest population (approximately 503,000). Manchester City Center is the city-region's primary cultural, business, and retail center. Population fell during the 1970s and 1980s but subsequently recovered, showing a growth rate of nearly 2% per annum over the period 2001–2011. The growth in population has not been equally spread across all age groups. In comparison to the national age profile, Manchester has a smaller proportion of residents aged 65 and older: 9.6% in Manchester compared with 16.9% in England and Wales (2011 figures). This is a result of natural losses and older people migrating to other areas in Greater Manchester. A study by the Audit Commission in 2008 found that: ". . . the older population . . . tends to be . . . poorer, isolated, and more vulnerable with a lower life expectancy and a need for acute interventions" (Audit Commission, 2008). The strong growth in Manchester's young and working-age populations over the last decade will have exacerbated this pattern (Bullen, Bendel, & McGarry, 2012). Research has also shown that older people in these environments face higher levels of disadvantage and social exclusion (Scharf, Phillipson, & Smith, 2003). While the probability of surviving to age 75 has slightly increased over the last decade, rates of survival in Manchester and Greater Manchester are still lower than those of England as a whole. The age-standardized mortality rate for all causes of death in 2008–2010 among people aged 65 to 74 years in Manchester (2,793 per 100,000) was 64% higher than that of England as a whole (1,703 per 100,000). Bullen et al. (2012) conclude that the characteristics of Manchester's older residents mean that they are more likely to place high demands on hospital emergency services and mental health services and experience long-term limiting illnesses at an earlier stage in their old age than seen nationally.

MANCHESTER AND BRUSSELS: PARTICIPATION IN THE WHO GLOBAL NETWORK OF AGE-FRIENDLY CITIES AND COMMUNITIES

A growing number of cities and communities worldwide are striving to develop new approaches to meeting the needs of their older residents. The WHO document (2007, p. 12) *Global Age-Friendly Cities: A Guide* outlines a framework for assessing the "age-friendliness" of a city and defines an age-friendly city as an "inclusive and accessible urban environment that promotes active aging." It identifies eight domains of urban life that might influence the quality of life of older people as shown in Table 1.1. Along with 11 other cities around the world, Brussels and Manchester were among the first cities[1] to be admitted into the WHO Global Network. Figure 1.1 shows the WHO Global Network five-year membership cycle, which includes planning, implementation, and evaluation (WHO, 2009). To join the WHO Global Network, Manchester and Brussels were committed to following the four steps in the planning phase: the establishment of mechanisms to involve older people in all stages of the age-friendly cities process; a comprehensive and inclusive baseline assessment of the age-friendliness of the city; development of a three-year citywide action plan based on assessment of findings; and identification of indicators to monitor progress against this plan.

The Involvement of Brussels in the WHO Global Network

In the case of Brussels, the immediate cause for the city council to apply for membership in the WHO Global Network was the city's participation in a research project called the Belgian Ageing Studies (BAS). This project

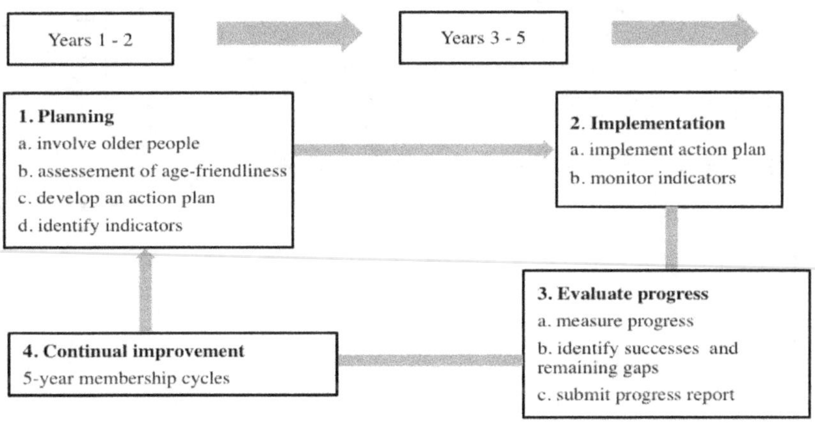

Figure 1.1 Cycle of World Health Organization Global Network of Age-Friendly Cities and Communities

Table 1.1 Frameworks of Age-Friendly Approaches by the WHO—Brussels and Manchester

WHO*	Brussels**	Manchester***
Outdoor spaces and buildings	Community safety • Organizing information sessions about crime prevention • Increasing feelings of safety through the presence of city guards (i.e., prevention agents) • Promoting cooperation between the police, the city council, court of justice, and local associations • Coordinated police actions in response to missing older people (c.f., project "senior focus")	Lifetime neighborhoods • Working toward accessible and well-designed living environments in which residents are not excluded by age • Working toward becoming an environmentally friendly Green City, with older people as important partners in achieving this Community safety • Preventing doorstep crime • Expanding initiatives to reduce crime and antisocial behavior on and around public transport
Housing	Housing • Ensuring that ordinary homes are suitable for older people • Promoting home adaptations and modifications • Supporting alternative housing options (e.g., group living schemes, intergenerational house-sharing)	Housing • Ensuring that ordinary homes are suitable for older people • Promoting Design for Access standards • Delivering additional new homes and affordable housing • Delivering new-care units
Transportation	Mobility • Making community transport more accessible and flexible including demand-responsive services • Ensuring a senior's lens in urban mobility plans • Supporting traffic-calming initiatives • Promoting existing car sharing and bike sharing initiatives • Improving the "walkability" of the city	Transport • Making community transport more accessible and flexible including demand-responsive services • Improving general bus services with better driving and more shelters with seats • "Bikeability" training for older workers • Improving road safety and pavements

Social participation	**Social life** • Improving information about activities and events • Offering a wide variety of activities (e.g., in local pavilions/meeting places for older people) • Promoting older people's participation in projects and actions concerning public life in the city (e.g., in district forums) • Supporting initiatives aimed at combating social isolation • Actively recruiting older volunteers • Organizing an annual "seniors week" with various activities	**Cross-cutting themes: Improving engagement; improving relationships** • Extending opportunities for older people's involvement in decision-making, project delivery, and service design • Expanding alliances of front-line staff, community groups, and older people • Extending opportunities for volunteering • Completing connected Generations Together programs and evaluating their impact • Reviewing best practice for a guide to tackle loneliness
Respect and social inclusion	• Actively involving the senior advisory board in all age-friendly projects and policies	**Cross-cutting theme: Promoting equality**
Civic participation and employment	**Socioeconomic situation** • Combating poverty through coordinated provision of social housing for older people • Promoting volunteering and engagement in associations • Recognizing the wide range of capacities and resources among older people	**Income and employment** • Support to over-50s to stay in work • Support to over-50s in steps toward the world of work (especially men who are made redundant and people on benefits due to mental health problems) • Promote and support volunteering and learning opportunities for over-50s

(*Continued*)

Table 1.1 (Continued)

WHO*	Brussels**	Manchester***
Communication and information	**Information** • Centralizing information regarding services, activities, and organizations for older people in a "senior guide" • Providing a bimonthly magazine offering an overview of leisure activities for older people • Facilitating information access (information sessions in senior centers and "pavilions"; Internet courses in libraries; easy access to/ discount purchases and rents of computers)	**Culture and learning** • Promoting more accessible and better marketed activities • Supporting community-based and intergenerational learning activities • Reaching minority ethnic elders and those from disadvantaged areas
Community support and health services	**Health** • Improving information about and access to available services • Better coordination of the fragmented supply of home care services	**Healthy aging** Care and support services • Developing intermediate care best practice • Introducing more integrated teams and multifactorial assessment • Improving home care and developing more comprehensive, flexible, and culturally sensitive services • Removing barriers to minority ethnic elders

Note. *For a more detailed description of the WHO framework and domains, see the document *Global Age-Friendly Cities: A Guide* (WHO, 2007).

**The analysis for the Brussels age-friendly city framework is based on internal documents that have been developed by the City Council on the basis of the research findings of the Belgian Ageing Studies project in order to be admitted into the WHO Global Network of Age-Friendly Cities and Communities.

***The analysis for the Manchester age-friendly city framework is based on the Manchester Ageing Strategy as developed in the policy document "Manchester: A Great Place to Grow Older 2010–2020" (see McGarry & Morris, 2011) addressed the first three steps of the planning phase in the WHO Global Network membership cycle (see Figure 1.1), using a participatory method involving older people to assess the age-friendliness of the city and developing a senior action plan based on an assessment of the findings. The BAS project is a research program that has been executed in over 120 municipalities in Belgium, with the aim of monitoring local challenges and opportunities, as well as issues of quality of life among home-dwelling people aged 60 and older. In 2008, the City Council of Brussels decided to participate in the BAS project in order to provide tools for evidenced-based age-friendly policies at the local level. The project was a result of a close collaboration among a research team, the city council, senior advisory board, local social services, and other stakeholders.

Drawing on a participatory methodology, older people were actively involved as actors at all stages of the project. The research was viewed as a tool for creating community networks, on the one hand, and empowerment of older people, on the other. The creation of a community network among the city council, the senior advisory board, members of local senior organizations, third-sector agencies, and other community stakeholders was crucial for developing broadly based support around the theme of age-friendly policies and environments. Empowerment was promoted through a cycle of research, participation, education, community action, and policymaking (Verté, Dury, De Witte, Buffel, & De Donder, 2009).

Key Initiatives in the Brussels Program

Following the completion of the BAS program in 2009, the councilor for senior affairs and her cabinet completed the WHO Global Network application showing that the BAS program succeeded in addressing all four planning criteria to join the WHO Global Network (see Figure 1.1). Similar to the WHO framework, the BAS program identified various domains of city life (community safety, housing, mobility, social life, socioeconomic development, information, and health) that might influence the quality of life of older people, as shown in Table 1.1, with a range of initiatives identified for each domain. In line with the city's aging policy/action plan, an important initiative from this work was the development of a joint publication between the mayor and elected councilors (Ville de Bruxelles, 2010), highlighting their commitment to the aging agenda in the city and describing some of the age-friendly initiatives to be undertaken by their respective departments. Given that "lack of information" was identified as a major problem by many older people who participated in the BAS research program, one of the initiatives concerned a range of actions to improve older people's access to information about services, social and cultural activities, etc., both through the distribution of a senior guide and through targeted information sessions for older people in senior centers, community centers, and local libraries.

Another initiative was developed by the department concerned with neighborhood renewal, introducing a range of interventions aimed at combating older people's fear of crime, an issue that was identified by older people as important in the BAS research project. In summary, the city's progress in terms of mainstreaming aging issues—or integrating the aging dimension within different policy fields to ensure that the needs of all age groups are met—might be considered one of the most important achievements to improve the age-friendliness of Brussels.

As a result of the above, the City of Brussels became the first Belgian city to join the WHO Global Network. A number of factors have determined the structure and focus of the Brussels program. It received *political and practical support* from elected councilors, particularly the councilor of senior affairs, as well as the senior service and the senior advisory board. Together they took a leading role in bringing together stakeholders who wanted innovative solutions to support an active and healthy aging and to improve the age-friendliness of the city. A second factor is the development of a "*citizenship*" perspective on engagement rather than a "*deficit model*" of aging: older people were leaders of the BAS project in Brussels, not passive participants. Third, as a result of the BAS research project, the council has developed insights into older residents' situations, as well as their needs and desires for aging well in the city. The nature of the BAS project made it possible to compare findings from Brussels with other cities that participated in the research,[2] highlighting a number of specific features of the older population in the city. Among others, these include a significantly higher percentage of elders (15.5%) experiencing "severe difficulties" in getting by on their incomes (compared with 6% in Flanders), a higher percentage of older people who rent in the private market (29.6% versus 8.28%) and social housing (17.7% versus 3.94%), and significantly higher levels of fear of crime and feelings of loneliness. At the same time, the research also demonstrates that, despite high levels of social exclusion experienced by older people in Brussels, the city also contains resources—such as close family ties, neighbors, and services—that combine to offer the potential for active and fulfilling lives (Verté et al., 2009). Such insights are crucial for identifying constraints and opportunities, both existing and potential, of the city for older people.

Manchester's Participation in the WHO Global Network

In the case of Manchester, work on age-friendly issues was linked to activities during the 1993 European Union Year of Older People and Solidarity Between Generations. The city council created a multi-departmental working group charged with promoting a broader range of opportunities and services for older people. The Manchester program was consolidated in 1998 with the establishment of the Better Government for Older People (BGOP) group.[3] In 2003, the Valuing Older People (VOP) partnership was launched, designed to accelerate work and develop partnerships with older people and a variety of organizations within the community. Between 2003 and 2010, the VOP program developed a variety of initiatives on the age-friendly theme. The first was an *engagement* program aimed at involving older residents in the leadership of VOP work. This involved setting up a representative older people's board, a wider forum of older people's groups, a community development program, and a small grants scheme aimed at developing these groups.

Key Initiatives Developed in the Manchester Program

The Manchester Ageing Strategy (MAS) was launched in 2009 following consultation with older residents, elected council members, and a panel of nationally recognized experts (Manchester City Council, 2009). The conceptual framework used by the MAS is similar to that developed by the WHO Global Network, with objectives focused on ensuring that older citizens will be more active and engaged, experience less inequality, receive better-quality care and support, and live in lifetime neighborhoods with flexible, affordable housing options. Under the auspices of the VOP program, the council and its partners have implemented initiatives to support the program. Other activities have focused on extending opportunities for older people's involvement in decision-making, project delivery, and service design.

As a consequence of the above, Manchester became the first UK city to join the WHO Global Network (in 2010), and work was consolidated with the official launch of the program in 2012. The VOP Workgroup consisted of staff from the National Health Service, local government, a housing trust, an arts agency, a national charity, and a local university. In 2014, the core team and broader program were relaunched under the Age-Friendly Manchester (AFM) banner, with a two-year development plan setting out action on four themes: age-friendly neighborhoods; knowledge and innovation; age-friendly services; and communication and innovation. A senior strategy group was also established, chaired by the Council's deputy leader, and a new action plan was developed based on the eight WHO domains.

Three main factors assisted the development of the program. *Political leadership*: Throughout the lifetime of the program, high-level political support was received from elected council members, principally the council's deputy leader with lead responsibility for equality strategies and the VOP lead councilor, a politician who has day-to-day contact with the VOP team. *A local narrative*: An important aspect of the work has been the development of an account of urban aging rooted in the city's disadvantaged communities, built around a combination of research findings and conceptual insights. Importantly, there are features of the city's older population that are atypical of older populations in most other local authority areas, including lower proportions and numbers of older people, a significant proportion of black and minority ethnic elders, higher levels of social exclusion and ill health, and high levels of population turnover. *Partnership strategy*: The city published two strategic statements creating a framework in which agencies can combine at citywide and neighborhood levels to achieve common objectives.

Similarities and Differences in the Age-Friendly Approaches of Brussels and Manchester

The conceptual frameworks used by the city councils of Brussels and Manchester are similar to that developed by the WHO Global Network. Table 1.1 compares the domains of the approaches and sets out the examples of initiatives being taken by Manchester and Brussels agencies in pursuit of age-friendly objectives. A number of similarities and differences between the approaches can be identified. First, both cities have dedicated teams of people influencing the delivery of policy directly or indirectly affecting older people. In Brussels, the "senior service" was established in 2007 with three members of staff, but reached a total staff of seven by 2013 dedicated to improving the quality of life of older people in the city. Manchester City Council's AFM team was formed in 2003 and works with council departments and organizations across the city to implement policies aimed at improving the age-friendliness of the city. It comprises five team members and five associate members (2013 figures). Both the senior service and the AFM are overseen by a board comprising older volunteers. From the outset, both teams sought to develop strategic partnerships and networks with the public sector and with voluntary and community organizations and collaborated with universities and researchers to assess the age-friendliness of the city. Both teams have also developed specific program and initiatives aimed at extending opportunities for older people's involvement in decision-making, project delivery, and service design.

A second similarity between age-friendly approaches in Brussels and Manchester is that the initiatives for social participation and inclusion (see Table 1.1) in both cities focus on extending opportunities for volunteering and involving older people in the development of age-friendly and intergenerational projects. In Brussels, the senior service team has developed interventions that aim to link the needs of younger and older people, for example, by involving older adults in guiding children in their school activities. Shared Places and Spaces is an example of a similar program in Manchester. This involves students from the Manchester School of Architecture who have examined the practicalities of using shared spaces that link younger and older residents (e.g., inner-city parks). Finally, similarities can also be noted in the theme of housing, with activity around expanding home improvement services, housing support services, and affordable housing schemes. Both cities aim to provide improved housing options with a better mix and choice of homes. This work was taken forward through the development of a practical guide aimed to raise awareness about the different housing options in Brussels (Makay & Lampaert, 2012) and through the council's strategic housing and social care group in the case of Manchester.

A number of differences can be identified between the cities. The first concerns the way in which aging programs have been "branded" and communicated. In the case of Brussels, the notion of "age-friendly Brussels" ("leeftijdsvriendelijk Brussel" in Dutch and "ville amie des aînés Bruxelles" in French) is not well known to the general public. Instead, older people tend to identify the city's aging program with the BAS project rather than with the term "age-friendly." The reason behind this is that the BAS project involved bottom-up participation of older people and thus implied a broadly based public support, whereas the application to become an age-friendly member of the WHO Global Network was a somewhat top-down initiative. Moreover, the age-friendly city framework is available only as an internal working document (in only one of the national languages), and there are few references to "age-friendly Brussels" on the city's website or at city public events. In contrast, the Manchester Ageing Strategy 2010–2020 is available on the city council's website and addresses the age-friendly domains using questions such as "Where we are now?", "Where we want to be?", "How we're going to get there?", and "Who's involved?" Moreover, the term "age-friendly Manchester" is explicitly used in public events and has become a brand or umbrella term for all initiatives in the city that aim to improve older people's quality of life.

A second contrast between the age-friendly city frameworks of Brussels and Manchester concerns the extent to which attention is paid to specific groups, such as minority ethnic groups and older people living in poverty. While Brussels is a highly ethnically diverse city with an increasing proportion of the older population foreign origin, there is less attention paid in the city's approach to developing policies on aging. The provision of social housing is seen as one of the main instruments to combat poverty, but, other than that, work seems to be largely focused on the average white older person. In contrast, the Manchester Ageing Strategy includes a number of targeted interventions and projects (especially in the domains of "promoting equality" and "healthy aging") aimed, for example, at "developing culturally sensitive services," "removing barriers to ethnic minority elders," and "reaching older people living in disadvantaged areas."

Third, a number of differences can be identified in relation to the age-friendly domain "outdoor spaces and buildings" in that the focus in Brussels is entirely on issues of community safety and crime prevention, whereas Manchester adopted a broader approach highlighting the importance of developing, initially, lifetime neighborhoods and, from 2012 onwards, age-friendly neighborhoods. A lifetime neighborhood is defined as a "place where a person's age doesn't affect their chances of having a good quality of life" with "services, infrastructure, housing, and public spaces that are designed to meet everyone's needs, regardless of how old they are"

(Manchester City Council, 2009, p. 37). This approach is being tested via a number of projects labeled as "age-friendly demonstrator" sites. In Brussels, the WHO domain of "outdoor spaces" is mainly covered by specific actions aimed to improve the community safety, an issue that was identified by older people as a main priority for Brussels in the BAS research project. "Senior focus" is one example of such a project that promotes cooperation and networking among the police, the city council, the court of justice, and community organizations to increase older people's feelings of safety as well as to develop coordinated police actions in response to missing older people.

DISCUSSION: BARRIERS TO IMPLEMENTING AGE-FRIENDLY POLICIES AND INITIATIVES

This chapter has reviewed age-friendly initiatives in Brussels and Manchester, as well as similarities and differences between the respective programs. Cultivating and maintaining the existing partnerships that the two cities have established and promoting broad-based collaboration with multiple stakeholders will be especially important to realize the potential benefit that the age-friendly initiatives have to offer. Another issue will be to further involve older people and those approaching old age as significant actors in setting the agenda for age-friendly developments. The fact that both cities have dedicated teams of people (the senior service in Brussels and the VOP team in Manchester) supporting varied initiatives and partnerships to improve older people's quality of life presents a major opportunity to reinforce a citywide age-friendly approach. Despite the opportunities, a number of barriers—both existing and potential—to implementing age-friendly programs can also be identified, including the prevalence of ageist attitudes and stereotypes; economic and political barriers; and potential limitations associated with the concept of "age-friendliness." The first of these may contribute to various forms of exclusion in later life and may undermine the work of age-friendly programs in promoting social participation. Older people may, for example, remain excluded from policymaking processes due to prejudices and stereotypes that label them "care-dependent." In Brussels, older people reported that while they have considerable influence over policies on *care-related issues* (such as home care, social services, and health care), they feel that they are often denied a voice on decisions concerning *housing issues, neighborhood design*, and *planning* (Vanmechelen et al., 2012). Similarly, evidence from Manchester (Scharf et al., 2003) suggests that urban regeneration policies could greatly benefit from the skills and experience of older people and the attachments they bring to their communities. Yet the evidence is that they often tend to be "invisible" in the implementation of such programs.

A second set of issues relates to economic and political barriers facing age-friendliness of cities. On the former, given a context of economic austerity, there are significant pressures to reduce funding for what might be called preventative programs. Taking the example of Brussels, there are recession-linked pressures in that the council has been unable to increase public spending on community health and social care in line with increasing demand. There is growing consensus that much more investment is required to improve support for carers and home helpers, to secure the availability of (semi) residential services (especially in densely populated and deprived urban districts), and to develop more flexible and culturally sensitive services in response to the challenges posed by increasing ethnic and economic diversity in the older population. A key issue will be how to increase investment in prevention to reduce *future* increases in demand when available budgets are under pressure to meet *existing* levels of demand. In the case of Manchester, plans to promote age-friendly neighborhoods may be compromised by budget cuts, which reduce public services such as libraries, information and advice centers, and day care facilities for older people. Threats to services may also lead to a public perception that the age-friendly brand is unrealistic and unlikely to be implemented given restrictions on public spending.

In terms of political barriers, Brussels and Manchester are cities experiencing a range of demands from their respective populations and aspirations to create "world class" environments that compete for investment with other national and international centers. The idea of age-friendliness thus has to compete with wider objectives associated with economic growth and development and, in consequence, may appear marginal to both. One of the major issues for both cities is the struggle for living space, not only between households but also between the wide range of functions of the city as an economic and educational center (and in the case of Brussels, as a diplomatic and political one). In this respect, cities need a more integrated strategy to achieve age-friendliness. One way forward could be to make age-friendliness a central part of policy-making, with a view to promoting sustainable urban development across its broad environmental, social, and economic dimensions. Successful implementation of such policies, however, will rely on the support of a range of stakeholders—including multiple levels of government; public, private, and third-sector organizations; and nongovernmental organizations. Reconciling the different interests and values of these groups will be a key issue for the age-friendly city movement to address.

Finally, the idea of age-friendliness itself carries limitations unless linked to wider debates about rights and citizenship within the urban environment. The concept of "the right to the city" is closely associated (as noted in the chapter's introduction) with the work of Lefebvre (1991) and has become a keyword for analyzing the struggles over the shape of the city and access to public space, or, in Harvey's terms (2009, p. 315), the right to "make and remake our cities and ourselves" under circumstances in which private capital is dominating the urban process. Purcell (2003) argues that "the right to the city" implies two main rights for its inhabitants. The first is to *appropriate* urban space; the right to "full and complete usage" of the city. The second concerns the right to *participate* centrally in decision-making surrounding the production of urban space. Such issues may be of particular importance for older people who become reliant on their immediate environment for achieving a fulfilling existence in old age. However, the so-called "paradox of neighborhood participation" (Buffel et al., 2012) applies especially well to older people; that is, they tend to spend a good deal of time in their neighborhoods (*being part of the city*), but are often among the last to be engaged when it comes to decision-making processes within their neighborhoods (*taking part in the city*). While cities are increasingly viewed as key drivers of a nation's economic and cultural success, their reconstruction is often to the detriment of those outside the labor market, especially those with low socioeconomic status. Achieving recognition of the needs of different generations within cities and exploiting the potential of the city for groups of all ages will be central to the process of making cities more age-friendly.

At the same time, the concept of "age-friendliness" must itself be kept under critical scrutiny, given the impact of economic austerity on many urban areas. Many of the cities in the WHO Global Network are themselves experiencing substantial reductions in physical infrastructure and services. The handling of these will be a major test for the ideals and initiatives associated with building supportive communities for older people. Given the challenges cities face, whether it makes a significant difference to the quality of people's lives to apply the age-friendly approach will need careful attention over the next phase of the movement's development.

CONCLUSIONS

The analysis in this chapter suggests that both cities would benefit from an exchange of experience and mutual learning. Promoting a dialogue between the city councils of Brussels and Manchester, together with the different stakeholder groups, could drive the age-friendly agenda forward in a number of ways. It could provide a repository of best practice cases, which would increase our understanding of what works and what does not work in particular policy areas. It could support the development and implementation of smart and innovative solutions to developing age-friendly communities. Brussels and Manchester would be able to share methodologies and indicators to monitor and evaluate the wider socioeconomic impact of investing in innovative services for living environments. Finally, providing a strong cross-national evidence base may also stimulate other cities to incorporate an explicit age-friendly focus into public policy.

The comparative analysis suggests that Manchester has a stronger tradition compared with Brussels in promoting age-friendly policies. This is especially the case in relation to Manchester's culturally sensitive and targeted approaches toward minority ethnic elders and those in deprived urban areas; the city's development of "lifetime and age-friendly neighborhoods," including the integration of aging-friendly principles into the regeneration and design of urban spaces; and the way in which Manchester is able to brand and promote the age-friendly approach. Manchester, on the other hand, has drawn on the experience of work such as the BAS. The development of BAS has been important in clarifying methodologies for working with diverse groups of older people living in urban communities. This work is now being partially replicated by the MAS, which has adopted a highly participative approach drawing on older people as co-researchers and interviewers (see also Bindels, Baur, Cox, Heijing, Abma, 2014)

Promoting the exchange of experience between both cities would offer a way forward in understanding key questions such as: "What are inclusive communities and how can we best achieve them?" and "How do individuals and communities most effectively make their voices heard?" These are especially relevant for aging populations, with older people often representing marginalized groups in urban contexts.

To conclude, both cities share a number of key success factors, which may prove to be helpful for other cities to develop age-friendly strategies. The first concerns an initial attempt to integrate policies for older people into those focusing on urban (re)development and the management of cities. Mainstreaming aging issues, a strategy of integrating aging issues into different policy fields, represents a major challenge for social and public policy as well as for community organizations and older people to ensure that the needs of all age groups are met in urban policy fields. A second factor relates to the focus on involving older people as significant actors in setting the agenda for age-friendly developments. A significant issue in this respect is the recognition that older adults are not just the beneficiaries of age-friendly communities; they have key roles to play in defining and fostering their distinctive features. As Menec et al. (2011, p. 24) have argued, implicit in the notion of age-friendly communities is the idea that older adults are an integral part of ensuring that a senior's lens is applied in decisions, policies, and planning: "Older adults must be part of identifying areas of need, prioritizing key issues, and ensuring appropriate implementation." Finally, both approaches have demonstrated the importance of building synergies and promoting partnerships with multiple stakeholders. This includes the involvement of public, private, and third-sector organizations and non-governmental organizations involved in developing age-friendly strategies. Within such joint efforts, leadership and coordination by local authorities have proved to be critical factors in developing age-friendly Manchester and Brussels.

Acknowledgments

The authors thank the team members of the senior service in Brussels, as well as the city councillor of senior affairs and her cabinet, for providing us with information and internal documents regarding the city's aging strategy.

The authors also thank the Age-Friendly Manchester City team members for their input and advice.

NOTES

1 The *cities* of Brussels and Manchester (not the Region of Brussels and Greater Manchester) applied for membership in the WHO Global Network.
2 More than 140 municipalities participated in the BAS project, the majority of which is located in the Flemish Region of Belgium. In each of these municipalities, a random sample from the population register was drawn, applying stratified quota where the proportion of features such as gender and age are identical to the underlying population (*N* = 64,737). Due to the participatory method involving older volunteers, the first response rate was between 65% and 85% in each of the municipalities (see www.belgianageingstudies.be).
3 BGOP was a government-initiated program of 32 "pilot" local government projects in the UK developing new approaches to engage older people within the community.

REFERENCES

Audit Commission. (2008). *Don't stop me now: Preparing for an ageing population.* London, UK: Audit Commission.

Bindels, J., Baur, V., Cox, K., Heijing, S., & Abma, T. (2014). Older people as co-researchers: A collaborative journey. *Ageing and Survey, 34*(6), 951–973.

Brussels-Capital Health and Social Observatory. (2008). *Poverty and ageing.* Brussels, Belgium: GGC. Available from http://www.observatbru.be/documents/graphics/rapport-pauvrete/rapport-pauvrete-2008-nl/2-armoede_verouderen_08.pdf

Buffel, T., & Phillipson, C. (2011). Experiences of place among older migrants living in inner-city neighbourhoods in Belgium and England. *Diversité urbaine, 11*(1), 13. doi:10.7202/1007742ar

Buffel, T., Phillipson, C., & Scharf, T. (2012). Ageing in urban environments: Developing "age-friendly" cities. *Critical Social Policy, 32*(4), 597–617.

Buffel, T., Phillipson, C., & Scharf, T. (2013). Experiences of neighbourhood exclusion and inclusion among older people living in deprived inner-city areas in Belgium and England [Special issue]. *Ageing & Society, 33*(1), 89–109. doi:10.1017/S0144686X12000542

Bullen, E., in collaboration with Bendel, N., & McGarry, P. (2012). *Older people in Manchester. A profile of residents aged 65 and older.* Manchester, UK: Manchester City Council.

De Spiegelaere, M., Closon, M.-C., Deboosere, P., & Humblet, P. (2009). Health and quality of life in Brussels. *Brussels Studies*, Synopsis 11. Available from http://www.brusselsstudies.be/en/brussels-studies-the-e-journal-for-academic-research-on-brussels

Deboosere, P. (2010). Brussels: A city, a region, a place to live. In *E-book 7: What does geography teach us about the future of Belgium's institutions?* (pp. 1–62). Brussels, Belgium: Re-Bel initiative.

Harvey, D. (2009). *Social justice in the city* (Rev. ed.). Athens, GA: University of Georgia Press.

Lefebvre, H. (1991). *The production of space.* Hoboken, NJ: Wiley-Blackwell.

Lui, C.-W., Everingham, J.-A., Warburton, J., Cuthill, M., & Bartlett, H. (2009). What makes a community age-friendly: A review of international literature. *Australasian Journal on Ageing*, *28*(3), 116–121. doi:10.1111/j.1741–6612.2009.00355.x

Makay, I., & Lampaert, L. (2012). *Wonen zonder zorg(en). 10 woonvormen om over na te denken*. Brussels, Belgium: Kenniscentrum Woonzorg Brussel.

Manchester City Council. (2009). *Manchester: A great place to grow older, 2010–2002*. Manchester, UK: Joint Health Unit.

McGarry, P., & Morris, J. (2011). A great place to grow older: A case study of how Manchester is developing an age-friendly city. *Working With Older People*, *15*(1), 38–46. doi:10.5042/wwop.2011.0119

Menec, V.H., Means, R., Keating, N., Parkhurst, G., & Eales, J. (2011). Conceptualizing age-friendly communities. *Canadian Journal on Aging/La Revue canadienne du vieillissement*, *30*(3), 479–493. doi:10.1017/S0714980811000237

Purcell, M. (2003). Citizenship and the right to the global city: Reimagining the capitalist world order. *International Journal of Urban and Regional Research*, *27*(3), 564–590. doi:10.1111/1468–2427.00467

Scharf, T., Phillipson, C., & Smith, A. (2003). Older people's perceptions of the neighbourhood: Evidence from socially deprived urban areas. *Sociological Research Online*, *8*(4). Available at http://www.socresonline.org.uk/8/4/scharf.html

Scharlach, A.E., & Lehning, A.J. (2013). Ageing-friendly communities and social inclusion in the United States of America [Special issue]. *Ageing & Society*, *33*(1), 110–136. doi:10.1017/S0144686X12000578

Vanmechelen, O., Verte, D., Teugels, H., Buffel, T., De Donder, L., Glorieux, M., . . . Verté, E. (2012). *Zorgnoden en-behoeften: de kijk van de Brusselaar. Analyse van sterktes, zwaktes, kansen en bedreigingen van de Brusselse woonzorg*. Brussels, Belgium: Kenniscentrum Woonzorg Brussel.

Verté, D., Dury, S., De Witte, N., Buffel, T., & De Donder, L. (2009). *Rapport van het ouderenbehoefteonderzoek Brussel. Monitor voor de stad en al zijn inwoners*. Brussels, Belgium: Vrije Universiteit Brussel.

Ville de Bruxelles. (2010). *Plan seniors 2010*. Brussels, Belgium: Ville de Bruxelles.

WHO. (2007). *Global age-friendly cities: A guide*. Geneva, Switzerland: World Health Organization.

WHO. (2009). *WHO Global network of age-friendly cities*. Geneva, Switzerland: World Health Organization.

2 Ireland's Age-Friendly Cities and Counties

The Development of a National Program

Sinead Shannon and Hugh O'Connor

INTRODUCTION AND OVERVIEW

While population aging has been on the global policy agenda for at least a decade, the aging of the population is only now becoming a policy issue in Ireland due to its differing demographic profile. In Ireland, the *Ageing Well Network*, a small nongovernmental organization established and supported by the philanthropic organization *Atlantic Philanthropies*, developed a national program of Age-Friendly Counties linked to the WHO Global Age-Friendly Cities program.

In the six years since its introduction in 2008, with limited resources, the WHO Age-Friendly Cities program has been adopted in 20 local authority areas, and a number of others are considering adoption. This chapter will trace the development from the early days of the program to the current day, in which the aim is to make every local authority in the country age-friendly.

Against the backdrop of a severe economic recession and cutbacks in both public-sector employment and other public resources, the program has succeeded in establishing both local and national structures and has reached the implementation phase in the majority of local authority areas in which it operates. The chapter will look at the background and policy context in which the program operates and will outline the structures that support the program at national and local levels. It will discuss a number of the key initiatives, some of which are small changes that make a big difference, while others are more systemic and long-term and effectively change the way that local service providers work together to make a difference in outcomes for older people. Finally, the chapter will look at the plans for evaluation at local levels and will highlight some of the challenges the program faces in making Ireland age-friendly.

Demographic Profile

Ireland currently has fewer older people and more young people than most other EU countries (Central Statistics Office [CSO], 2007) due to high levels of emigration in the 1950s (many of whom never returned), increased immigration during the 1990s, and two baby booms in the past 30 years, resulting in the highest percentage of children in the EU—at 25% of the total population.

These trends help explain Ireland's comparatively young population. However, this position is likely to change dramatically over the next decade. Predictions made recently by the central statistical office (CSO), based on a number of different scenarios (births, deaths, and migration rates), suggest that, under all scenarios, the number of older people is set to rise substantially. At the moment, the population aged under 15 (976,600) is considerably higher than the population aged over 65 (531,600), but it is predicted that this will reverse by 2031 (CSO, 2013).

Ireland has recently experienced an increase in overall life expectancy, which is expected to continue. Male life expectancy has increased by 20.5 years (since 1926) to 77.9 years in 2010, while females have seen a gain of 24.8 years (from 57.9 to 82.7). This trend is expected to continue, and male life expectancy is predicted to reach 85.1 in 2046, a gain of 7.2 years over the 30-year period, whereas for females there is a projected increase of 5.8 years to 88.5 (CSO, 2013).

In 2006, the proportion of older people stood at 11% of the total population, the lowest proportion of over-65s across all the EU's 27 countries. In the 2011 census, this proportion had increased to 11.7%. The number of people aged 65 years and older is projected to increase very significantly from its 2011 level of 532,000 to close to 1.4 million by 2046. The proportion of those older than 80 years is expected to rise even more dramatically. Over the next three decades, that proportion is expected to quadruple (increasing from 128,000 in 2011 to between 470,000 and 484,000 in 2046, depending on various scenarios) (CSO, 2013).

Key facts revealed by the 2011 census about older people in Ireland include the following:

- The rural population of almost all counties was older than in urban areas by almost two years (CSO, 2012).
- Home ownership is high in Ireland; in 2011, more than 87% of older people lived in owner-occupied accommodations (CSO, 2012).
- Almost 9% of older people (aged 65 and older) are still in employment, 14% of men and 4.6% of women (CSO, 2011a).
- Older people in Ireland are less likely (than younger people) to have access to the Internet. Almost six out of ten homes, where the head of household was aged 65 years and over, did not have any Internet access in 2011. This figure rose to 79% in cases where the person aged 65 or older lived alone (CSO, 2012).
- Car ownership is also relatively low among older people. In urban areas, almost one-third of people older than 65 years had no car, while in rural areas the figure was just under one-fifth (CSO, 2012).

CURRENT POLICY CONTEXT

As has been well documented elsewhere, since 2008 Ireland has undergone a significant economic adjustment, aimed at restoring the sustainability of the country's public finances. The main effect of this adjustment has been a dramatic cut in public expenditure through significant reductions in numbers working in the public service and an emphasis on public service reform. The approach to reform has frequently focused on the redesign of services, reconfiguration of resources, and greater use of volunteers.

There is a slowly growing recognition that Ireland, along with other countries, must begin now to respond to the challenge of demographic aging by promoting and supporting a positive experience of aging for all older people. Instead of later life being seen as a period of health decline and "retirement" from activity, it is increasingly regarded as a period in which older people are continuing to create and develop, a period in which they can enjoy their lives and actively contribute the experience and knowledge gained throughout their lifetimes to their families and communities.

The two previous Irish strategies relating to older people, both 20 years apart, focused primarily on health and care issues, and though the strategies committed to making important changes, many of these were not implemented. However, in the last few years, there have been a number of improvements in policymaking and provision for older people in Ireland, which can be seen as a significant momentum for change.

In fulfillment of commitments made in the then *Programme for Government*, a Minister for Older People was appointed in 2007 and an Office for Older People was established in 2008. The *Programme for Government* also committed to developing a strategy for older people, and the *National Positive Ageing Strategy* (NPAS) was finally published in 2013. The strategy provides a strong policy context for addressing issues relating to the quality of life of older people and, importantly, attaches particular priority to the importance of local structures supporting the implementation of the National Strategy. The Age-Friendly Cities and Counties (AFCC) program is cited in the National Strategy as an example of the type of structure that will contribute to the implementation of the strategy.

BACKGROUND

Following the recent enactment of legislation (Local Government Reform Act, 2014), local governments in Ireland have undergone a number of changes. There are now 31 local authorities, of which 26 are county councils, two are city *and* county councils, and three are city councils. Each council now has a chief executive, previously known as city or county manager, who is the manager of the local authority. Each county also has a number of elected representatives, depending on the size of the county's population (Environment, Community, and Local Government, 2014).

The members of the councils are the policymakers in local governments and are democratically elected in local elections held every five years. Most of the day-to-day management functions are carried out by the Ministerially appointed officials known as city or county managers (Local Government Management Agency, 2014).

The power of local authorities has been gradually eroding over the past few decades as they progressively lost control over services to national and regional bodies. For instance, since 1970, local government has had no health remit, and in the 1990s the National Roads Authority took overall authority for national roads projects, supported by local authorities who maintain the non-national roads system. The remaining responsibilities of the local authorities include housing; planning; roads and pavements; water supply and sewerage; environmental protection, including rivers, lakes, air, and noise; and amenities such as parks and public libraries. The city and town councils are responsible for the same services and functions but in more restricted geographical areas.

Following from an announcement in October 2012 by the Minister for the Environment, Community and Local Government, the abolition of all town councils and the merger of some county councils came into effect in 2014, to coincide with the next scheduled local elections (Department of Environment, 2014)

Unlike many other countries, public health services are delivered at local levels by the Health Service Executive (HSE), which was established by the Health Act, 2004, and came into official operation on January 1, 2005. The Minister for Health has overall responsibility for the HSE in government. These HSE areas provide and maintain hospitals, nursing homes, clinics, and health centers in their areas. Local authorities have no responsibility for the provision of local health or social care services.

HISTORY AND DEVELOPMENT

The Ageing Well Network, based in Dublin, was, up to December 2013, the host and policymaking structure for the Age-Friendly Cities and Counties program in Ireland. It was set up in 2007 as a "think tank," bringing together key leaders from government and non-government bodies and from the fields of policy development, academia, service delivery, and advocacy. The network was funded by the philanthropic organization *Atlantic Philanthropies* to develop and promote a new way of thinking about aging, focusing on the opportunities rather than the challenges.

In 2009, Ageing Well Network members saw the potential of the WHO Age-Friendly Cities Project to create a place where older people are empowered, enjoy good health, live securely, and continue to participate fully in society. For further information, see the introductory chapter in this volume for an overview of the WHO Age-Friendly Cities and Communities program.

The local authority manager in one county, Louth, was instrumental in bringing the Global Age-Friendly Cities Project to Ireland. Following participation in the WHO project, the Louth County Manager agreed with the Ageing Well Network to take the lead role in making Louth Ireland's first Age-Friendly County and in adopting an Age-Friendly Strategy, thereby launching phase one of the program in Louth in November 2008. Since then, the program has developed nationwide, with continued funding from Atlantic Philanthropies to provide a national manager, national coordinator, and five regional managers to guide and extend the program to all counties interested in participating.

The development of the Age-Friendly Cities and Counties program can be tailored to meet the needs, priorities, and preferences of older people in each individual city or county. However, all city and county programs share the following common features:

1. The Program is always embedded within existing resources and structures.
2. The Program finds ways to reconfigure the use of existing resources rather than creating new spends.
3. The Program adopts a collaborative and person-centered approach.

The program is now supported nationally by a team of seven—the Head of the Program, national coordinator, and five Regional Program Managers whose role is to support and work with counties/cities seeking to develop the initiative—while at the local level, local authorities provide varying levels of administrative and clerical support for the initiative.

Funding for the Ageing Well Network has now ceased, and the Age-Friendly Cities and Counties program is now managed by a new organization called "Age Friendly Ireland," located within the country's largest local authority, Dublin City Council. Soon, every Local Authority area in Ireland will have its own Age-Friendly Cities and Counties (AFCC) program.

Local Structures and the Development Processes

The success of the Program in Louth led to the development of a process to support its extension to other local authorities using the approach developed and tested in Louth. In the initial stages of development, the support of the local authority Manager is a crucial input to the establishment of an Age-Friendly Alliance. The Alliance then leads the program and is made up of the most senior decision-makers from the key public, private, and not-for-profit agencies. It is important that the Alliance is chaired by the local County Manager, particularly in the early stages of its development, and that it also has the support and involvement of key people from the local health service and the chairs of three additional bodies—the Older People's Council, the Service Providers, and Business Forums. The Alliance will also include key leaders from the police force (An Garda Siochana), third-level institutions, voluntary organizations, and other service providers in the area.

The Age-Friendly Alliance usually establishes a smaller Executive Group, which meets more frequently and has responsibility for managing the implementation of the Strategy, especially those projects involving collaborations of agencies. The Alliances focus on ways of ensuring that structures and services are adapted to be accessible to and inclusive of older people with varying needs and capabilities. In practice, this means that the work of the Alliance is to ensure that policies and programs that focus on the needs of older people become central to the work of the service providers.

Each city or county establishes an **Older People's Council** to ensure that the views of older people are central to the development of the program. These Councils are made up of representatives of all the older people's organizations in the county, and all older people in the area can participate through publicly organized meetings. These meetings give members the opportunity to determine key priorities for the Age-Friendly Strategy. In addition to representing the views of older people within the Alliance, the Council also sets up its own initiatives to address issues identified through a consultation process.

Another key structure is the **Business of Ageing Forum**, which works to stimulate awareness among the business community about how best to grow their customer base by deepening their understanding of older people's needs, preferences, behaviors, and attitudes. They are made up of business leaders who see the opportunities for businesses to benefit from their becoming more age-friendly.

The final core piece of the structure is the **Service Providers Forum**, which brings together all those organizations providing services to older people in the county—across the public, private, and not-for-profit sectors—with a view to exploring ways of improving the range and quality of those services and making them more responsive to the needs of older people. The forum seeks to identify ways in which services might be provided in a more cost-effective way, removing any areas of duplication and prioritizing areas of greatest needs.

Having launched the program publicly, the second phase of development begins with a comprehensive process of consultation throughout the city or county. A draft Strategic Plan is informed by the discussions and recommendations, and, once approved by the Alliance, the final Strategy is completed and launched. Having developed the Strategy, application is made to the World Health Organization for a WHO Certificate of Membership of the WHO Global Network of Age-Friendly Cities.

National Structures

At a national level, a number of structures support the implementation of the program. The ***National Implementation and Integration Group*** provides strategic support for the overall program and meets four times a year. This group is chaired by the Assistant Secretary General from the Prime Minister's Department (Department of An Taoiseach) and includes senior civil servants from the Older People's Office and other key government departments, representatives of the local authority County Managers, the police force, and the Executive Director and the AFCC National Program Manager from the Ageing Well Network.

The role of the group is to support the roll-out of the program across all Local Authority areas and to address any policy, structural, or other barriers that may be experienced. This group also examines ways of integrating and embedding the program into existing structures by aligning it with other initiatives, such as the children's policy area.

Expert Advisory Group

An Expert Advisory Group was established in 2012 to support and strengthen the program as it is rolled out nationally and over the longer-term. It is made up of people with strong backgrounds in relevant areas—environmental and social gerontology, transport, disability/access, older people's issues, safety/security, and the built environment. The group reviews the operation of the program and provides expert advice to both the National Implementation and Integration Group and the National and Regional Managers on ways of strengthening the program.

National NGO Forum

The Ageing Well Network also established a group that brings together representatives of the local nongovernmental organizations (NGOs) to explore and develop activities between their organizations and the AFCC program. NGO members of the Forum deliver a wide range of programs for older people, for example, advocacy, transport, information and communication, training, healthy living, and volunteering, all of which naturally tie in with the key areas of focus for the AFCC Program. The Forum serves a number of key functions:

- to support the NGOs as they extend their programs into new geographical areas;
- to facilitate partnerships between the AFCC Program and NGOs in local areas with a view to collaborating and possibly setting up new initiatives;

- to understand and respond better to the unmet needs of older people, as evidenced from the consultation processes and baseline studies conducted in participating counties; and
- to act as a catalyst in developing new initiatives in response to key areas of unmet need.

ACCOMPLISHMENTS—PROGRAMS AND ACTIVITIES

The Program is now fully operational in 19 local authority areas. Six further City/County Programs are set to launch during 2014. Exploratory discussions are also underway with representatives from a number of other cities or counties. Among the accomplishments of the Program to date has been the success in securing very senior-level engagement in each of the 19 participating local authority areas.

In the current economic climate, local authorities are under pressure to reduce expenditure without increasing income. In parallel to this, they must provide the same or even enhanced levels of services. This demand for more efficient and cost-effective delivery of services has created an impetus for change in the way that local authorities deliver services through greater cross-sectorial collaborations among central and local governments, the health sector, education, policing, and other parts of public service. The Program offers an opportunity to strengthen these links among and across these different sectors and to play a role in developing new and better ways to tackle complex societal goals.

Local Programs

The Age-Friendly Alliances have shown that strong local networks that promote flexibility, innovation, and greater local autonomy can develop more responsive, caring, and accessible services. At a time of greater competition for resources, the development of such cross-sectorial collaborations has created a culture that promotes greater integration of services and seeks joint solutions to problems, instead of confrontational relationships based on the protection of individual budgets.

One example of how the greater integration of services can bring about increased efficiency and cost savings is the "Fingal Health Route." The majority of mainstream public transport routes in Ireland (and particularly in Fingal, which is a local authority area north of Dublin city) are radial routes serving the needs of people travelling to and from Dublin city center. Direct public transport links to hospitals and other health services in the county are limited and in some cases non-existent. A considerable majority of older people across the county do not qualify for the Transport Service provided by the Health Service Executive. As a result, transport costs for older people to get to hospital appointments, particularly when travelling from the north of the county, can be extremely high. A survey of the region found that 47% of participants had missed their hospital consultant appointments (i.e., "no-shows") due to lack of transportation. Hospital appointments missed at short notice represent a considerable cost to the state.

In response to the needs of older people identified through the consultation process (undertaken as part of the Age-Friendly County Initiative), the local authority, Fingal County Council, in association with a local charity "Vantastic Ltd." developed a Pilot Door-to Door-Health Route for citizens over age 65. This is a free local Health Transport Service for older people living in North East and South East areas of Fingal. Over its first 12 months in operation, the Fingal Health Route Pilot helped 500 older people to attend health and medical appointments. This service is provided by a local service provider with charitable status. The pilot was developed as a collaborative venture between the local authority and the Department of Health.

An evaluation has recently been conducted that has identified cost savings of up to 50% in comparison with the traditionally tendered HSE (Health Service) Patient Transport Services. Informed by the evaluation findings, the HSE in that area has now agreed to reapportion their Public Transport Service budget to support the ongoing operation of the Health Route pilot. Following the publication of the evaluation, one of the local hospitals also agreed to coordinate and integrate its hospital appointments system in partnership with the Health Route. Reduced patient waiting times and timely appointments with clinicians are key outcomes secured through the introduction of this initiative.

Achieving political and senior policy level buy-in at the national level has been another success of the Age-Friendly Cities and Counties program in Ireland. In addition to receiving the support of senior civil servants, who participate in the National Implementation and Integration Group, the Minister for the Environment has also publicly endorsed the Program and has committed to extending it to all Local Authority areas across the country. Winning such enthusiastic governmental support is a distinct achievement.

On a local level, the AFCC Program has established a number of successful initiatives, often in collaboration with other organizations and service providers, to address particular needs within the area. Space constraints do not permit a full description of the various programs and activities taking place locally, but the following gives a brief outline of some. In one county, the development of the "Cúltaca" program supports frail older people who want to remain in their homes. This is an independent professional service that provides one-to-one community-based support to older people depending on their needs and circumstances. Currently in its fourth year, two Cúltaca posts are funded by the Health Service and Atlantic Philanthropies and serve a population of 1,000 older people.

In another area, the local hospital has introduced two new initiatives: a new code of practice for post-discharge arrangements for older people leaving the hospital and a new system of prioritization of older people at a hospital emergency unit, significantly shortening their waiting times for services. Auditing of outdoor spaces and buildings has been carried out in many areas and has resulted in more age-friendly footpaths, lighting, seating, bus shelters, and outdoor recreational areas, including the creation of outdoor Tone Zones and exercise machines, aqua aerobics, swimming, lessons, and tai chi for older people.

National Programs

On a national level, the Ageing Well Network (AWN) has also developed a number of interesting programs to address particular needs linked to the structures established through the AFCC Program.

Older People Remaining at Home (OPRAH)

In recognition of the desire of most older people to stay living in their own homes for as long as possible, the AWN developed a program aimed at providing evidence to support the implementation of a new, more personalized approach to the provision of care in the community. Supported by an Expert Advisory Working Group, the program has completed its first (development) phase and has produced recommendations on a new approach, which will be trialed over the next three years. The objectives of the initiative are:

- to enable frail older people to stay living in their own homes, where they might otherwise have been at risk of multiple hospitalizations or admission to long-term nursing home care;
- to reduce the number of hospital admissions; and
- to develop an action research project and, through evaluation, to develop evidence in support of a new way of providing health and social care.

To test the approach, it has been agreed to trial the initiative with 20 older adults participating in each of the six demonstrator sites located within the Age-Friendly City and County structures, with a view to an eventual National roll-out. The aim of the trials is to examine how an adequately supported home environment can be provided whereby:

- The person's social, care, health, and housing needs are assessed and a care plan developed in collaboration with them and their families to meet their particular needs, preferences, and priorities.
- The care is provided by formal and informal caregivers, and complemented with the use of technology, when and where appropriate.

- Families and the community have key roles to play in augmenting and enhancing the supports available through professional service organizations.

The position of a "Support Coordinator," to coordinate and manage the support package identified for each older adult, is central to the approach. Their role is to identify and assess the needs of the 20 participants, complete a holistic care plan, and imaginatively bring together the different supports they need to stay living at home. The Support Coordinator will draw on a broad holistic range of supports from the variety of public, voluntary, private, and community agencies in the area as informed by the individual older adult's needs and wishes.

Age-Friendly Towns

The Ageing Well Network has developed a program based on the experience of one age-friendly town: Ardee, Co Louth. In 2011, a spatial planning expert from the Dublin Institute of Technology (DIT) was recruited to engage and collaborate with local community leaders and other key stakeholders in improving the quality and experience of the town for older people.

In Ireland, every six years the local authorities develop Local Area Plans (LAPs) for towns in their areas. These LAPs are public statements of planning policies, prepared in consultation with the local communities and members of the public. The LAP sets out a strategy for the proper planning and sustainable development of an area, for example, village/town/district, and seeks to provide a framework for how a town can develop over time.

The revision of the LAP offered an opportunity for the creation of an Age-Friendly town. The Louth Age-Friendly County Program took the initiative and developed a submission for the Local Area Plan 2010–2016 in relation to Ardee. The submission by Louth Older People's Forum proposed amendments that would improve Ardee's potential to become an Age-Friendly Town. The submission was informed by the following:

1. A survey of existing areas and demographics within Ardee to determine the location, condition, and convenience of communities and the services and facilities most likely to be used by older populations.
2. A stakeholder consultation involving key informants in public authorities and agencies, as well as those currently involved in the provision of services to older people. A series of interviews with key stakeholders in Ardee took place in 2010. The interviewees were asked about their roles in and potential contributions to making Ardee an Age-Friendly Town.
3. Stakeholder consultation through a Community Consultation Workshop. This event captured the insights of older people in Ardee and

the views of the organizations that they represented. The participants were asked to identify what is good about Ardee and what is needed to make Ardee more age-friendly.

The real successes of the project were in the small changes that made a big difference to the community. For example, an access route was opened up to a retirement community parallel to the main street, saving a long circuitous walk. Traders agreed to give up two parking spaces on the main street, to allow for a bus stop in the center of the town, making it accessible for older people. Previously, the bus stop had been on the outskirts of the town, which was a long walk from the shops. The crossing time at the pedestrian lights was increased because there had been two fatalities of older pedestrians in the recent past. Benches already planned for the town were installed at suitable locations based on the input of older people, to facilitate walking to and from the center. Businesses took part in an Age-Friendly scheme, making small changes to improve the experience for older people; for example, the pharmacist introduced a delivery service for prescriptions.

The Age-Friendly Towns program has now recruited eight graduate planners located in 15 towns/areas to expand the program across the country. In stage 1 of this program, approval is sought for the process from the relevant county/city Age-Friendly Alliance, and, having received this, the planners recruit older volunteers to assist with the operation of the program.

In stage 2, the planners, with the assistance of the volunteers, carry out desk research, collect data, map the services of the town, carry out a walkability audit, survey older people, and carry out consultations with older people and other key players. Stage 3 sees the development of an action plan, a submission to the Local Area Plan, and implementation of the key actions, such as the establishment of a Wisdom time-bank.

EVALUATION OF OUTCOMES

As discussed earlier, Ireland's AFCC Program starts with a comprehensive consultation process in each county. The consultation is carried out through focus groups involving a wide representation of older peoples' groups, including minority and disability groups and those who experience rural isolation. The process also includes groups such as statutory agencies, intergenerational groups, and those providing services and supports for older people, along with business interests and arts/sport and cultural interest groups.

The process seeks to determine the priorities of older people in the area: what they feel is good in the county, what could improve, what they can do to improve things, and what they feel that others can do to make life better for older people in the county. The outcome of this consultation process forms the basis of the strategy developed for each city or county.

Evaluation

In some counties, the verbal consultation was accompanied by a baseline survey that was conducted at the start of the program. It was originally intended to be one of the core elements of the evaluation process. The aim was to establish the baseline position of older people at the start of the program and to measure progress towards goals by repeating the survey at various intervals throughout the implementation process.

To promote greater involvement among older people in the program, the AFCC regional manager in each area carrying out the baseline study recruited older people as volunteer researchers who were consulted on the content and involved in piloting administration of the survey. The Ageing Well Network supported the process by developing the instrument and providing training for the volunteers. The survey was administered over a three-week period by volunteers visiting randomly selected homes throughout the county.

Approximately 500 surveys were completed in two counties, and the volunteers were given modest stipends to cover travel costs. A professional survey company analyzed the data and produced an initial report. However, as more counties expressed interest in becoming age-friendly and the program expanded, it became clear that, even with the use of volunteers, the resources necessary to support the administration of a random sample study in up to 10 additional counties would not be available. Other less scientific approaches, such as Internet surveys or smaller sample sizes, were used in a number of additional counties by necessity.

Indicators

The second part of the evaluation process is the development of a series of indicators to measure the progress towards achievement of the goals of an Age-Friendly City or County. The aim is to collect data from service providers on the key factors that affect the position of older people living in the area. The process to develop indicators is currently ongoing and will be linked to the Ageing Well Network's outcomes-based planning approach and to the WHO process to develop core indicators to evaluate Age-Friendly Cities and Counties program.

This outcomes-based approach begins with the end in mind and then determines the actions and activities that will best achieve that end. By beginning the planning process with broad outcomes, it quickly makes it clear that no one agency on its own can improve these outcomes, but that progress requires various agencies to contribute and work collaboratively in a creative way to improve the lives of older people.

The approach also allows an Alliance to demonstrate progress in implementing its strategy—to show that it is making a difference—by identifying relevant ways of measuring progress and tracking these indicators over the lifetime of the strategy. This is an increasingly important consideration as resources become scarcer, and there's a more pressing need to account for their effective use. The outcomes, based on extensive consultations with older adults, are that older adults can:

- lead healthier and active lives for longer;
- stay living in their own homes and communities;
- get to where they need to go, when they need to;
- be enabled by the built and social environment;
- feel and be safe at home and out and about;
- have the information they need to live full lives;
- be truly valued and respected;
- participate in social, economic, and public life; and
- continue to learn, develop, and work.

CHALLENGES

Despite the recent accomplishments, Ireland faces some challenges in achieving its goal in making Ireland the best place in which to grow old. The current economic and fiscal climate has placed significant constraints on all agencies and has limited their abilities to engage in new initiatives that are time and resource intensive. In particular, the availability of suitably resourced and skilled Administrative Officers/Implementation Managers to support and coordinate the work of the local Alliances will be critical to the sustainability of the programs at local levels.

Interagency collaboration is challenging, complex, and time consuming. Building understanding, trust, and a willingness to engage cooperatively takes considerable energy and time. Where the process of group formation has to be repeated due to organizational change and the redeployment of senior managers, the complexity and timelines involved will be magnified.

The aging of our population has not, to date, been a political priority. Indeed, there has been a lack of awareness and exploration of the relevant policy issues. In some areas, stakeholders place increasing emphasis on the life-course approach, at the expense of a sole focus on older people. While this is likely to be both desirable and appropriate for the long term, it is somewhat premature given the need further to strengthen older people's voices at regional and national levels.

CONCLUSIONS

The Age-Friendly Cities and Counties program is now operational in many parts of the country. Its development will continue over the next few years until, it is hoped, all cities and counties in the country will have committed to becoming age-friendly.

Through its structures and the participation of older people, the program is helping to change the communities in which it exists so that older people can live healthy, engaged lives. The approach it uses is an example of how to plan and implement integrated services, "joined-up" thinking, and planning at local levels. This approach brings together the key agencies across public, private, and voluntary sectors to plan and integrate their services and supports, consulting with older people and enabling them to have roles in shaping how services are provided. It is a model in which the plans are developed and agreed upon locally, and the agencies involved agree to be accountable to each other at local levels.

There is a strong record of real change happening in areas such as housing, transport, information, representation, public spaces, and much more. Older people are represented in each Alliance through the Older People's Council. Initiatives such as the Age-Friendly Towns are working on the process of consultation, mapping, planning, and implementation of change projects on the ground. The Business Recognition Scheme encourages businesses to adapt their practices to acknowledge and meet the needs of older consumers. The Older People Remaining at Home (OPRAH) project is seeking to develop new ways of supporting frail older people who want to remain in their homes rather than enter nursing homes.

Locally, there is a well-developed process for getting the program established and the Ageing Well Network's regional managers support the program in its initial stages. The various Alliances have also built up strong track records of imaginative changes at local levels. One of the most significant achievements has been the development of interagency collaborations and partnerships that are a central part of the Age-Friendly Cities and Counties program. There is a very strong alignment between the Program's approach to interagency planning and the ambitions set out as part of government's Action Program for Effective Local Government: "Putting People First" (Department of the Environment, Community and Local Government, 2012). *Healthy Ireland*, the recently published health strategy, has also recognized the importance of partnerships in the creation of truly healthy communities. It states that, as part of the implementation of the strategy, "Relationships must be built between sectors to address the determinants of health and wellbeing" (Department of Health and Children, 2013).

That these developments have been achieved against the backdrop of Ireland's economic crisis is all the more remarkable. That so many local stakeholders have been prepared to show how outcomes for older people can be improved using current resources, mainly through better integration of service provision to offer improved efficiency and effectiveness, gives great hope that the program will achieve the vision of making Ireland the best place in which to grow old.

REFERENCES

Central Statistics Office. (2007). *Ageing in Ireland 2007*. Dublin: Central Statistics Office. Available from http://www.cso.ie/en/media/csoie/releasespublications/documents/otherreleases/2007/ageinginireland.pdf

Central Statistics Office (CSO). (2011a). *Statistical yearbook of Ireland 2011*. Dublin: Central Statistics Office.

Central Statistics Office (CSO). (2011b). *Survey on income and living conditions*. Thematic Report on the Elderly 2004 and 2009. Dublin: Central Statistics Office.

Central Statistics Office (CSO). (2012). *This is Ireland—highlights from census 2011*. Dublin: Central Statistics Office.

Central Statistics Office (CSO). (2013). *Population and labour force projections*. Dublin: Government of Ireland.

Department of the Environment, Community, and Local Government. (2012). *Putting people first—action programme for effective local government*. Dublin: Government of Ireland.

Department of Health and Children. (2013). *Healthy Ireland*. Dublin: Government of Ireland.

Environment, Community, and Local Government. (2014). *Local Government*. Available from http://www.environ.ie/en/LocalGovernment

Local Government Management Agency. (2014). *Local Government in Ireland*. Available from http://www.lgcsb.ie

Asia

3 From Age-Friendly Research to Age-Friendly City and Age-Friendly Regional Network

The Case of Tuymazy and Republic of Bashkortostan, Russian Federation[1]

Gulnara A. Minnigaleeva

INTRODUCTION

This chapter describes the development of age-friendly programs in the city of Tuymazy, Republic of Bashkortostan, Russian Federation. Currently there are 21 cities in the Republic of Bashkortostan participating in the WHO Age-Friendly Cities project. The program started in 2011 and has a more or less unified structure in all the participating cities, with a supervisory board at the Republic's level.

This chapter is a scientific analysis and case study, as well as a story of the author's personal journey. It is well known that many socially meaningful endeavors are started by people who have experienced or been close observers of some significant difficulties and have become anxious to help others overcome them. For the author, it was of utmost importance to help her retired mother to return to a meaningful and exciting life, which pensioners seemed to be deprived of due to various reasons at the beginning of the twenty-first century in Russia. Therefore, when the author had an opportunity during her internship at the World Health Organization (WHO) to work on planning of the Global Age-Friendly Cities project, she immediately decided to take it to Russia. It was clear that the scope of the project, its flexibility (the action plan targeted particular needs of a city and was based on available resources), and the WHO's "seal of approval" would make a relatively easy starting point for city officials towards improving the situations of older people.

The initial research stage was conducted in the capital city of Moscow and in the city of Tuymazy (hometown of the author).[2] However, the second phase took off only in Tuymazy because there was an interested party, or "champion" (Rogers, 2003), that is, the author and the nonprofit organization specifically created to promote the idea. After the first research phase was sponsored by the government of Canada, most efforts in project development were undertaken on a volunteer basis by the author, her mother,[3] and other dedicated volunteers who assumed different responsibilities at different stages of the projects. However, as the "older people" were climbing higher on the political agenda, the government of the Republic of Bashkortostan started supporting some projects within the scope of the age-friendly framework, including an international conference on aging in the Republic of Bashkortostan, third-age universities in every municipality, and other activities specifically targeting aging issues.

STARTING POINT: CONDITIONS THAT PRECEDED THE AGE-FRIENDLY CITIES PROGRAM

This section contains descriptive information on the area[4] and points out the problematic areas in the city of Tuymazy identified during the research phase of 2006 (conducted in accordance with the WHO Vancouver Research Protocol.)[5]

The municipality consisted of the city of Tuymazy and the Municipal Region of Tuymazy, with a total population of about 120,000 people. It is a typical industrial post-Soviet city surrounded by villages and rural communities in the municipal region. Among 66,300 residents of the city, the population over age 60 constituted 14% of the city population. Within this cohort, 5% were over 75 years old (2006). The municipality was located in the Republic of Bashkortostan, one of the administrative-territorial units of a sub-national division in the Russian Federation.

Life expectancy at the time was 59 for men and 72 for women.[6] Official pensionable age was 55 for women and 60 for men.[7] This age did not imply mandatory retirement.[8] Also, employees, having worked a certain number of years at particular jobs (such as military, secondary school teachers, miners, and quite a few others) were entitled to early retirement benefits provided by the government. Persons receiving retirement benefits were allowed to continue receiving salaries at the same time if they chose to continue working. The average Russian replacement ratio for pensions in 2006 was 27% (Monitoring, 2007), with the minimal pensions just above the poverty line (2,439 rubles [USD 90]). However, some groups, such as World War II veterans, were entitled to a pension three times higher.[9]

Whereas **housing and transportation** were generally perceived as good (residents were provided free apartments during the Soviet time and were allowed to privatize them at no cost afterwards), utility costs and disability-hostile design were major concerns. Establishing some kind of committee to monitor transportation and construction of new housing was a common suggestion. Some problems of the physical space arose from the unsolved social dilemmas; that is, community benches were occupied by youth gangs because they had nowhere else to gather.

Most **social services**, including healthcare, education, legal counselling, libraries, and services of social workers, were free and provided by the public sector; however, most of the focus group participants were not aware of the scope of services to which they were entitled. The topics of community support and health services were the most emotional and fraught of all. The topics of infringements and mistreatment in healthcare and unlawful extortion of payments in the city hospitals dominated discussions in most groups, with the exception of the public service provider group. Suggested solutions included introducing separate appointment lines and services open only to older persons and the creation of a community senior center with a variety of services.

Social participation of residents was provided by a number of public-sector facilities (libraries, recreation centers); however, none of them offered special programs for older people. Unlike in many Western countries, religious institutions did not play large roles in the community and engagement of older adults. Even though both Orthodox and Islamic (native to the area) religious traditions had been carefully revitalized in the population after 1991, only a portion of the population had readily embraced the renaissance of religion and cultural traditions of their ancestors, either Christian or Muslim, whereas many stuck to the atheist ideology with which they were brought up. Generally, culture and religion presented a twisted and complicated mix of the Soviet mentality, Russian (Slavic, 23.2% of population), Bashkir (33.4%), and Tatar (37.6%) customs, and Russian Orthodox and Islamic traditions, with many bilingual residents and both Russian and Bashkort (or Bashkir) being official languages in the area. This Soviet mentality, in fact, played an important role in the development of the age-friendly project afterwards. Due to the policy of absolute equality of different ethnicities in the Soviet times, there were no evident differences between lifestyles or social-economic status of different ethnic groups.

Social inclusion appeared to be a difficult issue, too. Overall, public attitudes towards old age were negative; however, older focus group participants reported unconditional respect from their close neighbors or relatives. People's awareness of senior citizens was heightened during the two major holidays that commemorate older persons' deeds; however, it remained dormant at other times. Opportunities to get involved and participate in social life were very limited. Suggestions for remedy included creating subsidized coffee shops for older persons, as well as more open clubs, and organizing intergenerational and educational opportunities.

Civic engagement was also quite problematic for many historical reasons. The voluntary sector was quite small and mostly oriented towards organizing leisure and social activities. Among a few hobby groups and children's clubs there were an association for persons with disabilities, an association for the blind, and veterans' organizations. The veterans' organizations were largely supported by the government, enjoyed benefits such as free rent of premises, some financial support, and participation in government's activities. Veterans there meant broadly any person retired from military service or from any other work. Automatically, any person who has reached retirement age became a member of a veterans' organization in his or her region.

No membership fees were required. The veterans' councils (boards) of these organizations were entities allowing their members participation in annual meetings and conferences. However, commonly they were not able to provide engagement for other retirees, who were considered members of the organizations. Opportunities for employment are extremely limited by open age discrimination at workplaces.

Communication and information channels included newspapers, television, and radio. Internet as a communication channel was starting to gain popularity but not among older adults, and access was still quite costly. The problem was with growing costs and clearly not enough effort to communicate relevant information with appropriate content for older persons.

BUILDING THE FOUNDATION: CONSIDERATIONS AND IDEAS FOR CONSISTENT WORK

After the research and evaluation phase of the focus group was completed in 2007, the most important task was to identify next steps. Overall, solutions on different topics proposed in the focus groups ranged from national solutions, such as raising pensions, to local people's collective efforts, such as organizing tree-planting projects. Evidently, some of the problems required action at a governmental level, such as the amount of pension allowance. However, many were potentially possible to be addressed at the city level. Moreover, citizens could potentially be organized to contribute to the solution of the problem. Some suggestions could be implemented by existing nonprofit organizations or by a new agency dedicated to senior citizen interests and organizing.[10]

Create a Champion

As suggested by Rogers (2003), famous people or influential officials, "champions" interested in innovative ideas, are needed to disseminate and implement the ideas quickly. A legally registered nonprofit, nongovernmental association created to be a "people's organization" could both serve as a champion for the idea and plainly perform some tasks identified during the research stage. As the existing organizations were not quite apt for the role, a decision was made to create a new organization. Even under conditions in which a nonprofit did not have substantial funding, volunteers could help to carry out many tasks, such as intergenerational programs and programs involving older persons in educating younger generation and peers; organizing third-age universities; organizing birthday and anniversary celebrations and other activities. Our small but committed team of only two people was legally able to register the non-for-profit association and start running it.

Most efforts had to continue to be voluntary because obtaining funding for NGOs in Russia was quite complicated, despite the fact that government spending on NGOs was steadily growing. In 2006, President Putin introduced an annual grant competition for NGOs; grants were provided on a competitive basis. The total grant pool increased from 1 billion rubles (about 37 million USD) in 2006 to 2.3 billion (about 70 million USD) in 2013. But many international foundations were squeezed out of the country by newly introduced regulations. Local authorities very rarely financially supported activities of NGOs (Institute of the Foundation of Public Opinion, 2001).

Incorporating the value and resourcefulness of older age into the name and mission of the organization was very important. After a few rounds of consultations with all the alliances we had gained by that time, it was decided to name the organization *Координационно—ресурсный центр для пенсионеров «Мои года—мое богатство»*, which literally translates into English as "Center for Coordinating Resources of/for Retired Persons. 'My Years Are My Treasure.'" We believe that this name allowed more senior citizens to identify themselves as a target audience of our organization. The legal name in English was chosen to be Organization of Retired Persons "Wisdom Ripening" (ORP WR).[11]

Building credibility was a challenge because, over the course of the economic transition of the 1990s, the Russian population was taught not to trust strange, new, or just unfamiliar structures, including nongovernmental organizations. In particular, charities and foundations were widely used for making gigantic profits and as instruments to establish "political capital" and were often abandoned by politicians as soon as they achieved their own goals (Korzh, 2007).

Our strategy of building the credibility consisted of the following steps:

1. "Borrowing" the credibility of a highly respected international organization, that is, the World Health Organization (WHO). Becoming a part of an international project based on its research and framework helped potential partners (and, first of all, city administration) to take us seriously and decide to lend their support.
2. Becoming "official," that is, becoming a legal body in accordance with the Russian nonprofit law. When initial contacts were made with potential partners and members, the first question was about the legal status of the organization.
3. Acting before asking for anything. Before we asked for any help from the city administration, we submitted a report on the research phase and a developed plan of the next steps. It proved that our intentions were serious, and we were not going to back off.
4. Being consistent. Every step in the plan we discussed was implemented and a report was given back to every person participating in organizing the activity.

Creating so called "free spaces" (Boyte, 2005) to encourage civic engagement was essential for success of the organization. Older people received "free spaces" where they could gather together to study, to celebrate, to realize smaller or bigger projects, or just to discuss everyday questions and have friendly chats. Before we could use the space granted by the Tuymazy College of Law and later by city administration, the founder's apartment served as a "free place" for gatherings of interested parties. The legal form of an association we chose also assisted with engaging its members (older persons) and providing participation opportunities.

However, organizing volunteer involvement was a challenge because of what we call "*post-Soviet 'allergy' towards volunteering.*" Civic participation in the Soviet Union was highly appreciated and encouraged. Yet it was mostly "mandatory volunteering" because people were in fact obliged to participate, which caused rejection and, in a way, "allergic" reactions. Approximately a quarter of the population participates in volunteer activities today (Mersiyanova & Korneyeva, 2011). Moreover, now we have a mix of individualistic lifestyles brought about by the new economic era and paternalistic expectations still persisting from the Soviet past. People still expect the government to provide all public goods. In 2002, more than 80% relied on the government (and not themselves) to solve all their problems (Mushtuk, 2003), including creating civil society.

ESTABLISHING PARTNERSHIPS WITH GOVERNMENT AND OTHER ORGANIZATIONS IN THE AREA

City Administration

Often, representatives of the government did not have a clear understanding of the nonprofit sector, its laws and reality, and its ways of improving the lives of older people (Minnigaleeva, 2011). However, 62% of NGOs somehow cooperated with local governmental institutions and city authorities, and only 26.3% had some relations at the level of regional governments of the Russian Federation (Baradachev, n.d.). Besides the WHO's approval of the project, the first successes of the organization were important to secure further support of the city administration, to establish connections with other organizations, and to have access to their non-monetary resources. The sense of ownership became so strong that, at every public occasion, city officials stressed that the project had started "with their help and participation."

Political Considerations

In the situation where any political competitors of the ruling party were being suppressed, our original intention of being apolitical proved to be the right choice. When one of the organization's members, on her own initiative, tried to work for another party for extra income before the local elections in 2009, it caused outrage among the city administration, and we narrowly avoided the stoppage of any support.

Other Organizations

There were also a few organizations that could potentially be used and be relatively available for cooperation in the city, including the network of veterans' organizations, with premises throughout the city, and a few state-subsidized clubs and recreational centers that could cooperate for organizing events, intergenerational activities, or the use of their premises. To recruit allies, we made an effort to learn more about other organizations and their leaders and to discover their self-interests through methods like "one on one interviews" (Alinsky, 1946). Involving their members and telling them about us was our intentional objective (the public work approach, Boyte, 2006), which led to some very productive partnerships.

Build Media and Public Relations

The role of the media is difficult to overlook in the modern world, yet nonprofit organizations in Russia were mostly invisible (CRCSNS, 2011). This is likely due to their neglect and lack of funds for representing nonprofits in the media. Setting up a website with a clearly identified mission and history, comprehensively structured descriptions of activities, and contact information was one of the first practical steps, which helped us to improve our visibility, served as proof of our trustworthiness, and allowed us to convey information about the ORP WR in a consistent and structured way. Also, the successes of the organization had to be visible to maintain a positive image (N. Kari, personal communication, 2007). Today, we have a more developed and frequently updated website in Russian (www.moigoda.org) and a brief version for our international partners (www.wisdomripening.org).

From the beginning, we partnered with media professionals at local levels who saw it in their self-interest to convey stories of civic engagement and older citizens (Register, 2000). The local newspaper (*Tuymazinsky Vestnik*) and local television have always been readily available to publish our stories and milestone events. We were also fortunate to have been picked for a federal-level news channel in the first year of the existence of the organization, which we believe contributed to our positive image and credibility.

Build on Culture

The importance of culture is central to effective public action of this kind. Success of developmental campaigns depends on understanding, respecting, and engaging cultural values and local traditions and mentalities (Rao &Walton, 2003). As stated earlier, the culture of older generations in Russia was shaped by so called "Soviet" values, mixed with traditions of ethnic or religious groups and the new capitalist ideology. The easiest way to start appeared to be to use a culturally familiar form with new content, such as a "People's University." This form of continuing education was widely used in the USSR; thus, it looked familiar and trustworthy for both older persons and the city administration. The city administration officials granted approval of the initiative and connected us with the management of a local college. All activities had to be free for older persons because of perceptions of their low incomes. In addition to our team's volunteer efforts, we needed in-kind resources: auditoriums for classes and volunteer teachers, which were provided by the college. And so the program began.

Start Small

One of the key elements in the "everyday politics of public work" framework is "starting small" (N. Kari, personal communication, 2007; Boyte, 2006) with a project that would be easy to do and therefore would develop more collective trust and confidence. As small projects get done, the people will want to do more, believe they can do more, and take on larger tasks. We started with two classes a week with 20 to 40 older adults participating. At first, just the author and a teacher of the local Law College volunteered to teach. The numbers tripled by the end of the school year, and, in just two years, the project grew into a long-term program with over 200 older students annually, 15 subjects taught, over 50 volunteers, and three other educational institutions permanently involved. Most of our teachers are older adults and volunteers; they first joined the association as members and later became interested in more active roles in the organization. Successful implementation of that first small project proved that we were capable and prepared to run the program, which attracted both partners and students and allowed us to start other programs, described below.

Recognize and Give Credit

Acknowledging contributions empowers and encourages contributors to continue to do more. Sharing the credit for success with partners is essential for keeping good relationships. Sharing the credit with our members is vital for their engagement and feelings of belonging and inclusion. Recognition is "the preventive for indignity and its many far-ranging consequences" (Fuller, 2006, p. 17). That really matters when we talk about senior citizens in Russia. They are discouraged that they are not being recognized in the society and that they are unable to contribute to the society. "Supply for recognition is unlimited," and we do our best to follow this rule (Fuller, 2006, p. 29) by acknowledging the efforts of our members and partners at public occasions, online, in publications, and in everyday work. One of our projects was to organize older people to provide recognition for other older people and engage them in organizing youth to participate in the same process as well.

MOVING FORWARD: FROM ONE AGE-FRIENDLY MUNICIPALITY OF TUYMAZY TO 21 AGE-FRIENDLY CITIES IN THE REGIONAL PROGRAM IN BASHKORTOSTAN

Policy Structure

In 2010, the head of Administration of the Tuymazy Municipality signed the application letter to become a part of the AFC Network. Even though the mayor was supportive of the program, it was ORP "Wisdom Ripening" (ORP WR) that was expected to play the major role in implementing age-friendly ideas. The city watched the start of the AFC Network in the Republic before adopting a bylaw regulating AFC in Tuymazy. The structure of the regulating body was suggested by the ORP WR and accepted by the city administration. A working group was created for each of the WHO domains, with each group involving both city administration representatives and older persons who were members of "Wisdom Ripening." Specific bylaws have been created to legitimize the policy structure and make it more capable of solving the issues. Unfortunately, securing funding for paid staff turned out to be an impossible task.

In 2011, the President of the Republic of Bashkortostan approved participation of the Republic in the WHO Global AFC Network. The responsibility of administering the program was initially delegated to the Association of Municipalities. The approach to the project here was "top-down" and to a large extent required cities to participate, which is typical for many large-scale initiatives in the Russian Federation. The policy structure from top down consists of 1) the regional supervisory council; 2) executive body, which consists of three entities (Association of Municipalities of the Republic of Bashkortostan; "Wisdom Ripening"; Ministry of Labor and Social Welfare); and 3) coordinating councils in each participating city.

Even though the regional program gained strong political support, no funding was received for administrative purposes. The implementation of the program's procedures largely relied on the volunteer efforts of Association of Municipalities' staff, Bashkortostan's Ministry of Labor and Social Welfare, or ORP WR, all already overloaded with other tasks. Naturally, that leads to instability in the work flow.

Planning and Reporting

As required by the WHO recommendations, Bashkortostan's participants have gone through the process of evaluation, planning, and reporting on the initiatives in their respective cities. A schedule was planned so that cities would have to collect information and assemble reports during times when normally they would not have other pertinent engagements. However, human resource shortages make it almost impossible to make full use of the reporting system for control and improvement of city programs.

Building Awareness

The economic and political changes that followed the collapse of 1991 drastically changed the traditions of respect and care for the older generations, which had been deeply rooted in the patriarchic society of the tsar and Soviet Russia. After 1991, suddenly older people became the only reminders of the previous epoch. Many of them had difficulties getting used to the new economic reality and to having less security and more freedom. The perceptions that older persons were the poorest, most vulnerable, and least apt for the new social order turned out to be difficult to change 20 years after the collapse; most political leaders and social authorities were still thinking of older people as care recipients and a social burden (Minnigaleeva, Zulkarnaev, Demina, KHizhnjakova, Sabitova (Jusupova), 2013). The strategy of expanding awareness included organizing special education and age-friendly events and facilitating participation of interested parties, with high-ranking officials at the events promoting age-friendly policies. It helped the officials to see a bigger picture and realize that aging was becoming an important item on the international and national policy agendas. We made special efforts to involve key political figures in such events as 1) at the national level, the "50-Plus Forum in Moscow"; 2) at the regional level, the "Conference 50-Plus: All Positives of Ageing" in Ufa; 3) seminars for deputy heads of municipalities on social issues; 4) and, locally, "World Cafes"[12] (Brown & Isaacs, 2005) in some municipalities (Minnigaleeva, 2011).

Programs Established

The age-friendly cities idea is an overarching concept that may incorporate different solutions targeting the same goal of making cities livable and enjoyable for people at any age. Today, programs in the city of Tuymazy are largely based on the projects championed by ORP WR. For example, there are two third-age universities (as of 2013), which developed from the first experiences of ORP WR. ORP WR organizes courses on the basis of demand and the interests of members (according to member survey results). Local schools and colleges allow the use of their facilities, when free from regular classes, at no cost. Such a strategy is also beneficial for young students; it makes them observe and incorporate active aging.

The TV Studio "50+" project was significant for raising awareness and spreading age-friendly ideas. In 2007, it was started on a volunteer basis by Elena Kornilova, the chief editor of the local television company, with our support. It has become a very helpful tool in communicating ideas important for older persons. Ms. Kornilova trains older students and helps them to produce shows, which are on air monthly. The shows mainly target the older population, but quite often raises issues of significant public concern, such as environmental problems or community life.

The results of focus group research suggested that there was a big gap between the cultures of the younger and older generations. "How can we sew them together?" was the key question.[13] We created opportunities for intergenerational activities, communication, and interaction. Over the course of two years, about 30 college students and 10 high school students became volunteer tutors for older learners of computer skills. As our survey showed, after a course (12 classes during 3 months), 30% to 50% of the students changed their opinions about older persons and gained a better view of them as "active, willing, and able to learn and develop." Some older adults also realized that "the youngsters were not gangsters; they could be very patient and caring."

Civic engagement was made a cornerstone in the foundation of the ORP WR. The project "Grandmothers for Children" invites older women volunteers to work with children with disabilities or from disadvantaged families. Grandmothers volunteer to help 6–10-year-olds study, read, and draw or just spend time with them. The professional support of psychologists or social workers is provided if needed. "Grandmothers for Grandmothers" is based on volunteer practices by AgeConcern Ukraine. Older women visit others who are yet older, live alone, and are not healthy enough to go outside or become members of our organization. These visits do not require funds, but rather time and dedication from older volunteers.

The idea of travel clubs for senior people is one that is difficult to believe in for many Russians because of the image of older people as passive and poor (Minnigaleeva, Zulkarnaev, Demina, KHizhnjakova, Sabitova (Jusupova), 2012). However, ORP WR members started organizing trips, negotiated discounts with travel companies, and are happily enjoying budget travelling. The trips range from just out of the city to intercity and abroad.

When there are dedicated and interested individuals in the government (like Bashkortostan's current Minister of Labor and Social Welfare, Lenara Ivanova), they learn, follow, and find ways to provide government funding for similar programs. For example, two very popular regional programs for older adults were adopted directly, based on the practices of ORP WR. *People's Universities of Third Age* is a regional initiative, sponsored by the government of the Republic of Bashkortostan. It provides funding to establish programs for older adults. The subjects were defined in accordance with the data from surveys of older people in different cities. From 2011–2013, 19.6 million rubles were allocated to finance the program Republic-wide, and more than 40 thousand retired persons took courses through the program.[14] *Social tourism* programs provide older people with opportunities for travelling within the Russian Federation at a significant discount; the regional government covers 70% of the cost of travel within Bashkortostan and 30% of the cost of travel within Russia. According to information on the official program's website, 2,988 older persons and persons with disabilities participated in the program.[15] Some municipalities (for example, the city of Tuymazy) follow suit and provide local matching funds for local programs of social tourism. Other regional programs in the age-friendly domain include: 1) barrier-free environments; 2) provision of dental and certain other health services, which are otherwise not readily available through the state healthcare system because of additional cost, or because of waiting lines; 3) local municipalities' initiatives. The Barrier-Free Environment program is a federal-level program, but each region develops its own implementation plan and reports results to Moscow.

The most important pieces are, perhaps, initiatives of local municipalities, which started independently in accordance with the age-friendly framework and have the potential to be adopted by others. For example, in the city of Ufa (capital of Bashkortostan), learners at the third-age university, as a result of one of our World Cafes, were thrilled with the idea of organizing a *service center*, where cloth making, repair services, or hairstyling could be provided for older adults by older adults on a free basis or at a discount. With some lobbying efforts on their part, and some negotiations and planning, a service center was opened, backed by government funding. Another promising idea implemented in the city of Ufa is a Third-Age University Alumni Association. In this city with a population over one million and third-age universities operating in every district, an alumni network may serve as a basis for new projects and more engagement of older people. A beautiful project in the area of intergenerational solidarity and connectedness has been undertaken in the city of Sibay. Here they run a *"Family Tree" ("Shezherai")* project, for which young and old together research their histories and origins, draw family trees, and then talk about their experiences at a citywide celebration. The celebration involves traditional dances, food, and clothes. Such events indeed help to connect the generations and celebrate culture and the rich heritage of the region.

CHALLENGES AND CONSTRAINTS

Besides the challenges nonprofit sectors generally face in Russia, such as mistrust by the general public, difficulties of obtaining funding, "allergies" to volunteering, and low rates of civic engagement, there are some specific constraints with which we are still struggling.

Language Barrier

All official age-friendly documentation is developed and published in English, as well as age-friendly research and practices. That means that all the literature has to be translated to Russian before we can use it for our purposes, which, of course, makes the process longer, more expensive, and less efficient. We hope that introduction of the new website for the WHO age-friendly networks (www.agefriendlyworld.org) will help to mitigate the problem by making multiple languages available.

Communications and Technology

Older adults are the driving force in the AFC project; however, communication technology for many of them (especially those older than 65) still constitutes a challenge in Russia. Most older people, retired from jobs several years ago without having used computers at their jobs, are only starting to learn to use them for personal reasons. Therefore, we have to provide all the communication and information face to face, via telephone, in print, or in public meetings, which brings additional costs and time constraints. Computer skills courses for older adults are very popular today, which may improve the situation in the future.

CONCLUSIONS

The Global Age-Friendly Project of the World Health Organization (WHO), enriched by civic engagement theory and practice, offers a wonderful basis for improving older people's situations. This project of the highly respected international body offers credibility, a verified methodology, and possibilities for international support and exchange.

In our experience, a small grassroots initiative supported by a well thought through strategy has led to the development of a large regional program. The key suggestions we have followed include creating an agency to trigger, promote, and implement age-friendly practices; establishing partnerships with government and other organizations in the area; building and maintaining media and public relations; building on culture; expanding and encouraging the civic engagement of our members; starting small and growing; and providing recognition and credit.

Special attention should be paid to building and maintaining the credibility of the organization; incorporating the values and resourcefulness of older ages in the mission, name, and all of the organization's activities; and creating "free spaces" to encourage the civic engagement of the members who have become a major driving force and resource for the development of the organization. To expand and develop, every effort should be made to build awareness of age-friendly ideas and the WHO global project and to create partnerships with dedicated individuals and organizations embracing the idea of a society for all ages. With multiple stakeholders involved, it is essential to maintain regular communications and stick to the planning and reporting schedule. For continuity and sustainability of a large-scale project, it is essential to hire paid staff.

Acknowledgments

This chapter is an output of a research project implemented as a part of the Basic Research Program at the National Research University Higher School of Economics (HSE).

NOTES

1 This chapter is an output of a research project implemented as part of the Basic Research Program at the National Research University Higher School of Economics (HSE). I deeply appreciate help, guidance, and editorial support from: *Louise Plouffe*, whose approval and trust made it possible to conduct the research in Tuymazy, which later led to the development of age-friendly cities in Bashkortostan; *Harry Boyte*, who helped me to work through the framework of developing age-friendly cities in Bashkortostan after the initial research; *Jan Hively*, who has become a role model and a source of encouragement and inspiration to continue my work on aging and community building.

2 We are deeply grateful to my supervisor Louise Plouffe, Alexandre Kalache, director of the Ageing and Life Course Programme at WHO in 2006, and the Government of Canada, which sponsored the initial research phase in many cities, including Moscow and Tuymazy.

3 I deeply appreciate unconditional support and understanding from my family: Zakira Minnigaleeva, Afruz Minnigaleev, Yulia Yusupova, and my lovely nieces.

4 Hereafter the community profile data are based on Gareyev, A. M., Gataullin, R. F., Mukhametshina, L. M., Geography and ecology of Tuymazinskiy region. 2005, 174 p. or as provided in interviews by the city government employees.

5 The focus group research was conducted in full accordance with *Vancouver Research Protocol*. Available online at http://www.who.int/ageing/publications/Microsoft%20Word%20-%20AFC_Vancouver_protocol.pdf. Accessed 19.09.2013. In total, 11 focus group interviews were conducted in August–November 2006. There were eight groups with older persons (four groups with persons 60–74 years old, and four groups with persons of 75 and older); two groups with caregivers; and a series of three focus groups conducted with service providers. These data have limitations drawn from the nature of the process as well as from the research circumstances.

6 Life expectancy increased to 64 for men and 76 for women in 2012.

7 There is ongoing political discussion about increasing the retirement age; however, this idea is strongly opposed by the public and has not changed as of 2013.

8 Government employees were required to retire from their positions at the retirement (pensionable) age until 2011.

9 Minimum (or so called "social") pension provided by the government to every person who has achieved retirement age (55 for women and 60 for men) increased to about 230 USD in 2012. Also, the so called "three pillar" pension system based on solidarity system plus personal and employer contributions was enforced. Thus, some retirees today receive much higher pensions than the minimal.

10 It is worth mentioning that, unless it is stated otherwise in the text, the work described herein was fulfilled on volunteer bases with the in-kind support by some organizations (see text) and volunteer contributions of the people.

11 We are grateful to Almira Fattakhova and Jan Hively who suggested, respectively, the Russian and the English titles.

12 World café is a method of community organizing based on collective wisdom, communal interests, and use of conversational process in a relaxed café-like atmosphere (Brown & Isaacs, 2005). The method was recommended to us by Jan Hively and Moira Allan, who have used it for their *Pass it on network, dedicated to sharing ideas on good practices for older persons*. See also their website at http://passitonnetwork.org/

13 Conclusions were based on focus group research and interviews conducted during the author's coursework on the "Civic Engagement" class taught by Harry Boyte in 2007 at the Humphrey Institute of Public Affairs, University of Minnesota.

14 The Program of People's Universities of the Third Age and analytical data are available online at http://mintrudrb.ru/retiree_university

15 Information is available online at the Social tourism program's website: http://xn—90alcrhmdckk5a.xn—p1ai/about.php

REFERENCES

Alinsky, S. (1946). *Reveille for radicals*. Chicago: University of Chicago Press.

Baradachev, I. (n.d.). Obshhestvennye iniciativy i vzaimodejstvie nekommercheskih organizacij i organov vlasti v regionah Sibiri: opyt realizacii social'nyh proektov MOF SCPOI [Public Initiatives and Cooperation of Non-Profit Organizations with Government Bodies in Siberia: the Experience of the IRPF SCISC Social Projects]. Available from http://www.prof.msu.ru/publ/book6/c61_02.htm

Boyte, H. (2005, September/October). Reframing democracy: Governance, civic agency, and politics. *Public Administration Review*, *65*(5), 537.

Boyte, H. (2006). Citizenship as public work. In J. Tulchin & M. Ruthenberg (Eds.), *Citizenship in Latin America* (Chapter 14). Washington: Woodrow Wilson Center.

Brown, J., & Isaacs, D. (2005). *The World Café: Shaping our futures through conversations that matter*. San Francisco, CA: Berrett-Koehler Publishers.

Centre for Research on Civil Society and Nonprofit Sector [CRCSNS], National Research University Higher School of Economics [NRU HSE]. (2011). *Annual monitoring of civil society and the nonprofit sector*. Unpublished raw data.

Fuller, R.W. (2006). *All rise: Somebodies, nobodies, and the politics of dignity*. San Francisco, CA: Berrett-Koehler.

Gareyev, A. M., Gataullin, R. F., & Mukhametshina, L. M. (2005). *Geography and ecology of Tuymazinskiy region*. 174 pp.

Institute of the Foundation of Public Opinion. (2001). Obshhestvennye organizacii v Rossii. Pogovorim o grazhdanskom obshhestve [NGOs in Russia. Let's talk about civic society]. Available from http://lib.socio.msu.ru/l/library?e=d-000-00---001ucheb--00-0-0-0prompt-10---4------0-0l--1-ru-50---20-help---00031-001-1-0windowsZz-1251-10&a=d&cl=CL1&d=HASH01e83b843f09090b33e76b36.4

Korzh, B. (2007, March 7). Nekommercheskie organizacii. Doverjaj—proverjaja [Nonprofit organizations. Trust and check]. *Online newspaper "Nashe Vremya."* Available from http://www.raso.ru/?action=show&id=13162

Mersiyanova, I., & Korneyeva, I. (2011). Vovlechennost' naselenija v neformal'nye praktiki grazhdanskogo obshhestva i dejatel'nost' NKOZ: regional'noe izmerenie [Involvement of population in the practices of civil society and NGO activities: Regional dimensions]. Moscow: NRU HSE. 196 pp.

Minnigaleeva, G. (2011). *The role of nonprofit organizations in creating conditions for self-realization of older adults at municipal levels*. Unpublished manuscript. Moscow: National Research University Higher School of Economics.

Minnigaleeva, G., Zulkarnaev, T., Demina, K., KHizhnjakova, I., & Sabitova (Jusupova), L. (2012). Pozhiloj chelovek v glazah drugih pokolenij: znachenie dlja obshhestva i lichnosti [Older persons in the eyes of other generations: implications for the society and individuals]. *Vse pljusy zrelogo vozrasta, 1*, 90–99.

Minnigaleeva, G., Zulkarnaev, T., Demina, K., KHizhnjakova, I., & Sabitova (Jusupova), L. (2013). Obraz pozhilykh lyudey segodnya [The image of older people today]. *Psikhologiya zrelosti i stareniya, 4*, 1–24.

Monitoring. Newspaper "Kommersant" No 8. (2007, January 24), p. 2. Available from http://www.kommersant.ru/doc/736448

Mushtuk, O. (2003). Otnoshenie moskvichej k dejatel'nosti v stolice obshhestvennyh organizacij kak pervoosnovy formirovanija grazhdanskogo obshhestva [Attitudes of the Moscovites towards activities of NGOs in the capital as the basis for building a civil society]. *E-journal Pul's,* 2(274). Available from http://www.mos.ru/cgi-bin/pbl_web?vid=1&osn_id=0&subr_unom=2109&datedoc=0

Rao, V., & Walton, M. (2003). *Culture and public action: How cultural factors affect an unequal world.* Stanford, CA: Stanford University Press.

Register, C. (2000). *Packinghouse daughter: A memoir.* St. Paul: Minnesota Historical Society, 288 pp.

Rogers, E. M. (2003). *Diffusion of innovations.* New York: Free Press, 551 pp.

4 Successful Aging in a High-Density City-State

A Review of Singapore's Aging Policies and Urban Initiatives

Keng Hua Chong, Zheng Jia, Debbie Loo, and Mihye Cho

INTRODUCTION

Singapore is experiencing a tremendous shift in the age of its population. For over 30 years, the overall fertility rate in Singapore has been lower than the replacement rate of 2.1 children born to a woman.[1] At the same time, the average life expectancy in Singapore has increased from 66 years in 1970 to 82 years in 2010 (NPTD, 2013, pp. 9–10). A turning point was reached in 2012, as the first cohort of the Baby Boomer generation[2] turned 65 years old. In 2013, there were 404,400 residents aged 65 years and above, which was about 10.5% of the total resident population of 3,844,800 (DOS, 2013).[3] It is expected that by 2030, more than 900,000 residents (about 21% of the projected resident population) will be over age 65 and will retire from the workforce (NPTD, 2013, p. 12).[4] The old-age support ratio[5] is also projected to decline from 5.9 in 2013 to 2.1 in 2030 (meaning that for every elder/retired person in 2030, there will be 2.1 working-aged persons to support them) (NPDT, 2013, pp. 12–13). This unprecedented aging of the population will bring significant challenges to the small and dense city-state of Singapore.

Aging has been part of Singapore's national agenda since the 1980s. Recognizing that aging issues are complex and multifaceted, the Singapore government has consistently used a multi-agency approach, forming several "high-level" inter-ministerial committees to monitor aging trends, propose policies, and coordinate efforts among the various governmental agencies. Key thrusts of these committees' focus have included housing and land use, accessibility, healthcare, eldercare, financial security, employability, lifestyles, and well-being.

Singapore's aging policies are guided by a vision termed "Successful Aging." This concept was first introduced by the Inter-Ministerial Committee on the Ageing Population (IMC) in 1999, and later further developed by the Committee on Ageing Issues (CAI) in 2006. The vision of "Successful Aging" is defined by processes rather than by goals; and it is described in four distinct levels: the "individual level," the "family level," the "community level," and the "national level." The vision states that to achieve "Successful Aging," individuals should "assume personal responsibility and start preparing for old age early." Families are "the first line of support" and should take care of their elder members. Communities should offer "a range of comprehensive services to support families." And the government is recommended to provide "necessary policies and programmes to empower Singaporeans to lead meaningful lives in their golden years." Thus, this vision indicates that all sectors of society should contribute to this collective will (CAI, 2006, p. ii; IMC, 1999, p. 10).

Singapore's aging policies are intrinsically linked with its urban development. The unique urban setting of Singapore lies in its geographical constraint as a small island of 715 km^2, its high population density of over 7,400 people per square kilometer, and its scarcity of natural resources (DOS, 2013). Decades of centralized urban planning by governmental agencies have created a high-density and transit-oriented urban environment, guided by the principle of optimizing land use and economic potential. The various activities of senior citizens (i.e., living, transportation, healthcare, and recreational and social activities) all take place within this highly planned urban environment. Since the 1960s, the Housing & Development Board (HDB) has developed high-rise public housing estates, locally called HDB estates, which are hallmarks of the urban landscape in Singapore. These HDB estates also house 82% of the population, including most elder Singaporeans (DOS, 2013). The 1995 "National Survey of Senior Citizens in Singapore" discovered that awareness of various community-based health services was generally low among elders.[6] It is evident that, since then, subsequent initiatives by the state have been conducted to administer more widespread provisioning of community-based healthcare support, and to move towards more community-centered planning through concerted urban initiatives that range from improving senior citizens' housing environs and their accessibility to providing various types of care and social services in their neighborhoods.

This chapter examines Singapore's aging policies by reviewing the state's urban development initiatives through the lens of "Successful Aging." It begins with an analysis of the development of various aging policies—their principles and underlying values—over the years. With a larger and more diverse aging population, the government has been again looking into some of these policies and proposing new urban initiatives to address aging issues in the recent years. As such, we will base our review on the four strategic thrusts laid out in the 2006 CAI Report (i.e., elder-friendly housing, barrier-free accessibility, integrated healthcare and eldercare, opportunities for active lifestyle) and evaluate recent implementations and planning by various agencies since 2006. Though other aspects such as financial security and employability are equally important, this chapter focuses only on issues related to the urban environment. A separate section discusses civic participation in urban development, an emerging demand as the senior population becomes more involved.

DEVELOPMENT OF SINGAPORE'S AGING POLICIES

Over the years, each of the inter-ministerial committees on aging has responded to contemporary issues and anticipated future needs. During this time, aging policies in Singapore have consistently emphasized the family as the first line of support in elder care—this in accordance with Singapore's "minimalist approach to social welfare" (Mehta & Briscoe, 2004) and with justification based on an Asian culture of "filial piety" (Mehta, 2006, p. 45). Fostering filial piety became the main approach towards caring for the aged since the formation of the 1982 Committee on the Problems of the Aged led by the Ministry of Health (MOH, 1984, p. 41), which also viewed placing elders in institutions as a "last resort" and an "extreme measure" (MOH, 1984, pp. 16, 34).

The early stage of the process of developing aging policies coincided with rapid urban transformation, which had resulted in large-scale resettlement of a population from low-rise shophouses, row-houses, and village houses to high-rise housing units, constituting the further breakdown of previously close-knit large families and neighborhoods (MOH, 1984, p. 11). Measures that help to foster family ties and neighborliness in order to take care of elders, such as encouraging aged parents to live with or close to their children, thus became important in this new urban living environment.

The 1995 National Survey of Senior Citizens in Singapore Commissioned by Ministry of Community Development (MCD)[7] gathered information examining the amenities used and places visited by senior citizens, their modes of transport, and their awareness and use of community-based support services (MCD, 1995). These findings were translated into the Report of the Inter-Ministerial Committee (IMC) on the Ageing Population in 1999, which called for collective efforts from the four levels of society—individual, family, community, and state—to realize the vision of "Successful Aging," that is, to ensure that all levels of society are well prepared for the challenges and opportunities of an aging Singapore (IMC, 1999, p. 2). The report proposed that government agencies should create and implement more "vehicles" to aid community members as they age. The report also introduced the key principle of "aging-in-place"[8] along with strategies such as elder-friendly homes and environments, various housing options, integrated community planning, and the provision of healthcare and social services for elders by Volunteer Welfare Organizations (VWOs) (IMC, 1999, pp. 10–12). The 1999 IMC report established a holistic vision of "Successful Aging," which has subsequently been further developed in later aging policies, with an underlying philosophy that the state should only "help families and the community play their role, not supplant them" (IMC, 1999, p. 21).

Nevertheless, stronger emphasis had been placed on community planning, as seen in the Eldercare Masterplan 2001–2005, released in 2001 by Ministry of Community Development and Sports (MCDS). In this Masterplan, community-based facilities for elders were proposed to be "planned and built as an infrastructure that is required for the functioning of the community, just as markets, schools, and hospitals are part and parcel of the national infrastructure," with a recognition of the need for adequate support services from the community, as "the family is under increasing stress" (MCDS, 2001, p. 7). This eldercare Masterplan covered physical infrastructure, community service systems, funding policies, programs for both elders and caregivers, and public education, with an underlying approach of "Many Helping Hands" that included the community, families, and senior citizens themselves in dealing with aging issues (MCDS, 2001, pp. 8–27). This multi-thronged approach was then used to guide community planning and urban development in the subsequent decade.

The 2006 Report on Ageing Issues by the Committee on Ageing Issues (CAI) provided a comprehensive five-year Masterplan with recognition of the significant transformations in the demographics of Singapore—most notably, the looming retirement of the baby boom generation. The largest cohort of seniors, the baby boomers, started reaching age 65 in 2012. The CAI Report placed a new and urgent emphasis on maintaining a high quality of life for this cohort, as it would be healthier, much more educated, and more economically well off than earlier generations (CAI, 2006, p. i). Though it reaffirmed the fundamental role that family should continue to play in supporting elders, the report offered concrete policy recommendations to be implemented by multiple government agencies in four strategic thrusts: housing for seniors (elder-friendly housing), accessibility for seniors (barrier-free society), caring for seniors (holistic and affordable healthcare and eldercare), and opportunities for seniors (active lifestyles and well-being) (CAI, 2006, pp. iii–v).

In response to the issues cited in the CAI report, a high-level Ministerial Committee on Ageing (MCA) was set up in 2007 to coordinate the various national efforts. A report on the State of the Elderly in Singapore was subsequently published in 2009. Recognizing the contribution of seniors, "participation" was added as one of the key pillars of the aging policy framework (along with the more conventional pillars of "health" and "finance"). The report also adopted the 2007 World Health Organization's (WHO 2007) concept of "active aging," which became one of the four new strategic thrusts of the aging policy, along with aging-in-place, healthcare & eldercare, and employment & financial security (MCA, 2009, pp. 3–5).

By the early 2010s, several changes in the demographics of the elder population became apparent. First, due to changing social norms and family structures, there were more seniors living alone (or with each other) than with their own families (MCA, 2012). Second, seniors are now living longer, and also with more chronic health issues. At the same time, the healthcare model had also shifted from a "cure at all cost" model to "managing a gradual decline with an emphasis on well-being and happiness." This healthcare model demanded more step-down care[9] and a preference for even more home-based services, given the scarcity of land in Singapore (TODAY, 2013). Third, a better-educated and information-rich public began to demand more involvement in decision-making and community planning. When it comes to planning elder facilities, the issue could become more complex due to competing land use in a high-density environment.[10] "Participation" and "Inclusive Environment" thus become key drivers in aging policies or initiatives that deal with urban development, as reflected in a recent speech by the Health Minister:

> On our part, the government will engage advisors, grassroots leaders and residents on our plans for facility development and to strike a careful balance between addressing the needs of the seniors and the needs of residents in the vicinity. We need the support of grassroots organizations and Singaporeans themselves to make the community inclusive and welcoming for our seniors. (Speech by Minister Gan Kim Yong on MCA Initiatives, 2012)

Since the establishment of the vision of "Successful Aging" in the 1999 IMC Report and the introduction of the "Many Helping Hands" approach in the 2001 Eldercare Masterplan, emphasis has been placed on individual responsibility and family support as much as on community-based eldercare facilities and other urban infrastructures to enable aging-in-place. Besides the benefits of staying put in the community with which elders are familiar, the strategy of aging-in-place also brings other advantages, such as space-saving and relatively minimal intervention for a land-scarce, high-density city-state. Aging policy and urban development are thus intrinsically linked in Singapore. In particular, the 2006 CAI Report has clearly laid out four concrete strategies to guide the development of urban policies and initiatives by various government agencies. These four strategic thrusts (elder-friendly housing, barrier-free accessibility, integrated healthcare and eldercare, opportunities for active lifestyle) hence serve as an overall framework on which we will evaluate the recent urban initiatives and plans from 2006 to 2013 in terms of their extents, effectiveness, and gaps in meeting the needs of seniors.

OVERVIEW OF URBAN INITIATIVES IN RESPONSE TO AGING ISSUES

Housing for Seniors: Elder-Friendly Housing

The vision of housing for seniors is to make Singapore "an elder-friendly place, one that allows and encourages older persons to live as part of the family and community," with enhancing quality of life and enabling aging-in-place as key concerns (CAI, 2006, p. 16).

The CAI recommended that government agencies provide seniors a range of housing options with elder-friendly features and also assist them in monetizing their housing assets. The main measures included offering more rental options for seniors, facilitating the development of retirement housing by the private sector, continuing housing upgrade programs to make rental blocks and old estates more accessible and age-friendly, providing support services in the community, and providing reverse mortgage schemes and options for seniors to "downgrade" the existing flats, etc. (CAI, 2006, pp. 18–22).

A variety of housing options were initiated by the Housing & Development Board (HDB) to cater to the needs of elders, including Studio Apartments (SAs), one-room rental flats for those in the lower-income bracket, and the recently launched 3-Generation (3Gen) flats[11] for multigenerational families.

Since 1998, HDB has introduced the ready-to-occupy SAs that come with built-in wardrobes, kitchen cabinets, and various elder-friendly features to meet the needs of elder residents.[12] They are sold on 30-year leases, and only those above age 55 are eligible to purchase these units for their own residences. Available in two sizes targeted at one- or two-member households, they are usually envisioned to house aged singles or elder couples. These SAs have since been introduced island-wide, often within housing estates with well-established transportation networks, easily accessible facilities and amenities, where they exist in either standalone blocks or are integrated with larger flat types.

While the elder-friendly home features are generally welcomed by the seniors, the nature of SA tends to result in particularly age-segregated communities. A study in 2008 measured the "quality of life" of elders living in SAs, and uncovered that, despite the evident benefits of elder-friendly home features and provision of senior-centric services and facilities in these "conducive studio apartment environments," the benefits were out-weighed by the "corresponding negative effect of the psychological and social environments, which result from the monotonic social group that breeds morbidity, a sense of rejection and despair" (Addae-Dappah, 2008, p. 152). The findings of this study highlighted that about 74% of elders in these high-rise studio apartments actually preferred age-integration to age-segregation, and 78% considered age segregation detrimental to their "personal well-being." Yet, despite the study, another 4800 SA units have been planned for construction over the coming two to three years (The Sunday Times, 2012). Currently, SA is the major housing typology specifically designed and constructed by HDB for the large group of middle-income elders. Although SAs have effectively offered affordable and age-friendly flats for elders living by themselves, it is worthwhile to investigate further the social and psychological impacts of current planning and design of SAs, as well as to explore other housing alternatives for this group of elders before launching massive SA projects in a short period of time.

Moving into SAs is one of the various methods HDB recommends to seniors who want to downgrade and monetize their current homes. Other schemes include the so-called “Silver Housing Bonus” and the “Enhanced Lease Buyback” (The Sunday Times, 2013). The former allows the senior homeowner to sell his/her flat and downsize to a smaller one, while receiving a cash bonus. The latter scheme allows the homeowner to sell a portion of their existing flat lease back to HDB for some retirement income, while continuing to live in the flat for the next 30 years. This second scheme supports the notion of aging-in-place. However, for those who choose to downgrade from their existing flats, this could mean moving away from their current familiar neighborhoods. Doing this at the late age of 55 and older, it can be difficult for seniors to adjust to the new community and neighbors, which exacerbates a sense of isolation and loneliness. The increased promotion of recent downgrading schemes, while aiming to help elders to monetize their larger flats, may also contribute to an increased estrangement and disrupted emotional connections to their existing urban and social networks that were formed over a larger part of their middle age.

On the level of housing policy, the Married Child Priority Scheme (MCPS) encourages married children and parents to reside together, or within the same estate or neighboring estate to foster care of aged parents. The Multi-Generation Priority Scheme (MGPS) encourages families to stay close to each other in specifically integrated flat types. These are envisioned to promote multigenerational living and strengthen extended familial networks.

Aside from financial programs, HDB has also undertaken a widespread initiative to enhance the built environment for its residents, with special attention on assisting elders to age in place comfortably and conveniently. HDB introduced upgrading schemes that aimed to tackle not only the issues faced by a growing population of seniors in the public housing estates, but also the issue of aging buildings themselves. These schemes fall under the Main Upgrading Programme (MUP) introduced in 1990, which was renamed Housing Improvement Programme (HIP) in 2007, continuing the efforts to enhance the overall living environment of housing estates. Improvement works have been undertaken at the precinct level, the block level, and the unit level. Particular works at the precinct level include improvements and additions of covered linkways between blocks and reconstruction of existing pedestrian footpaths, stairs, and ramps for accessibility. Improvements at the block level include improvements of lift lobbies and upgrading of lifts to stop at every floor, where feasible. Unit-level improvements include upgrading existing toilets with grab bars and adding new pedestal pans. One of the most significant and highly visible schemes was the highly subsidized nationwide Lift Upgrading Programme (LUP), which was launched in 2001 and concluded in late 2014 (The Business Times, 2013). This program seeks to upgrade existing lifts and lift shafts and to provide new lift shafts for flats built before 1990. The goal is to provide direct lift access at every floor to benefit elders, families with very young children, and the disabled. Other schemes include the 2007 Neighbourhood Renewal Programme (NRP), which aims for a comprehensive integration of improvements across various neighboring compounds. Residents can contribute by suggesting and proposing the types of improvements they would like to have through this scheme. This represents an attempt to incorporate more active consultation and participation from the community on the improvements to be provided in their housing precincts.

More recent schemes have been specifically introduced to focus on age-friendly features and support aging-in-place through unit- and block-level upgrading and enhancements. These include Project Lift Improvement and Facilities Enhancement for Elderly (Project LIFE) and Enhancement for Active Seniors (EASE) in 2012. Project LIFE is particularly targeted at one-room rental flats, whereby an "Alert Alarm System" (AAS) is installed in the bathroom and living room in each unit. When assistance is required, the elder resident can pull an emergency cord, triggering an alarm that alerts the relevant personnel or, in the case of emergency, the Civil Defense unit.

Most of the elder-friendly physical upgrading schemes have been prioritized for smaller dwelling types (such as one- and two-room flats) and have tended to benefit lower-income elder residents. Results from the 2011 National Survey found that a large majority of dwelling types still do not include elder-friendly features such as non-slip tiles and grab bars in bathrooms, which are part of the upgrade provisions under the Project EASE scheme targeting dwelling types of three rooms or smaller. Larger flats that may not yet benefit from these schemes also have elder residents who may require these elder-friendly features. Presently, only 26% of all dwelling types contain elder-friendly home fixtures (MCYS, 2011). HDB still faces upcoming challenges in catering to the needs of an increasing population of senior citizens who live in its public housing estates nationwide.

Accessibility for Seniors: Barrier-Free Society

The vision for accessibility for seniors is to make Singapore "an inclusive elder-friendly place, one that allows older persons to integrate with the wider community and lead active lives," providing "safe and unhindered access" in homes, buildings, and the outdoor built environment, as well as a "user-friendly" transport system (CAI, 2006, p. 23). To achieve an inclusive, elder-friendly "Accessible City" for all, the CAI suggested coordinating efforts by major government agencies such as the Building and Construction Authority (BCA) and the Land Transport Authority (LTA), as well as by the network of public housing estates managed by the Town Councils island-wide. The main goal of accessibility requires improving barrier-free access between these estates and transport networks.

Major recommendations put forth by the CAI included: requiring LTA and private transport companies to provide public buses with "low-floor, step-free and wheelchair-accessible" entrances; improving barrier-free features for pedestrians on roads; providing education and promotion of universal design through courses and guidelines; improving inter-connectivity between buildings; legislating the compliance and maintenance of the Code on Barrier-Free Accessibility; and establishing inter-agency committees on barrier-free accessibility (CAI, 2006, pp. 25–32). This section presents a review of the major initiatives by BCA and LTA implemented after 2006.

BCA's 2007 Code on Accessibility was an improvement over the previous "2002 Code on Barrier-Free Accessibility in Buildings," which had undergone three reviews since its 1990 inception. The 2007 Code was developed through a review committee comprising stakeholders and the public, including inter-agency representatives, industry players, and voluntary welfare organizations, whose feedback was sought.

One of the key improvements of the 2007 Code over its 2002 predecessor was the inclusion of "new accessibility features to link buildings to buildings, as well as from buildings to parks, open spaces, transport nodes and bus shelters" (BCA, 2007). It was also new to support "aging-in-place" by stipulating elder-friendly features within residential units that can be retrofitted in future, including design of corridors and doorways, the installation heights of switches and sockets, and the provision of lifts to make all floors and communal areas accessible. Furthermore, the 2007 Code provided guidelines for better barrier-free design with improved interconnectivity of parks and open spaces, transport stations and nodes like bus shelters, shopping complexes, and industrial buildings.

While these guidelines provide a minimum checklist for designing and ensuring a universally accessible built environment, barrier-free improvements to existing buildings need to be better integrated with their urban context. Some barrier-free features such as handicap ramps were designed and built with excessively long winders to mediate differences in levels between buildings and pathways. These well-intentioned elder-friendly features may result in seniors having to walk longer distances. More sensibly designed and innovative urban features are needed to enhance accessibility.

To achieve further the vision of "an inclusive elder-friendly place," BCA released its latest update of the Accessibility Code in 2013 after a month-long public consultation on the draft. The new code was envisioned to support government's efforts in building an inclusive society, to make it accessible for people with diverse needs, to offer a better quality of life not just for persons with disabilities, families with young children, and elders, but also for everyone in the society, who may be caregivers to those with special needs (BCA, 2013). Some of the significant improvements include a more integrated method of designing barrier-free access to buildings through holistic consideration of various types of building arrivals (e.g., connectivity with taxi shelters, handicap car park lots); provisions of sanitary facilities that meet diverse needs (e.g., accessible individual washrooms for persons with disabilities accompanied by caregivers of the opposite sex); provisions of wheelchair-accessible transfer space; guidelines of stair design for ambulant disabled people; and more inclusive design guidelines for space usage, facilities, and elements within buildings.

It is evident that BCA's Code on Accessibility has been improving to become more inclusive and integrative. It has been realized that barrier-free design is integral to the design of buildings, including the need to provide more integrated and innovative links and interconnectivity between different parts of public spaces and the larger urban environment.

While BCA is enforcing barrier-free design in built environments, LTA is striving to enhance the transport system to make it more user-friendly for elder users. In 2008, LTA launched the Land Transport Masterplan, with a vision of moving towards "a more people-centered land transport system that will meet the diverse needs of an inclusive, livable, and vibrant global city" (LTA, 2008, p. 4). Meeting diverse needs is among the strategic thrusts identified,[13] under which several urban initiatives beneficial to seniors were proposed, such as expanding barrier-free facilities, assisting low-income groups, facilitating cycling, and engaging communities (LTA, 2008, pp. 70–81, 89–90).

Significant improvements of the land transport system have been achieved since 2008. For example, more low-floor, wheelchair-accessible buses were introduced; all Mass Rapid Transit (MRT) stations and bus interchanges now have barrier-free access pathways; accessibility of walkways, bus stops, and taxi stands was improved; shelters were installed on pedestrian overhead bridges; and lifts were installed at overhead bridges next to some MRT stations. Besides physical infrastructure, transit fares became more flexible based on distance and thus more affordable, especially for low-income elders. A special "Green Man Plus" scheme[14] was also test-bedded to allow elders and disabled pedestrians to have extended time while crossing the roads (LTA, 2013, p. 6–12).

With the recognition of higher expectations of commuters, a growing population and economy, as well as tighter land constraints with more difficult trade-offs,[15] LTA released Land Transport Master Plan (2013) to improve further the land transport system. The latest public consultation revealed that the public valued a more comprehensive public transport network, better quality of MRT and bus services, and more comfortable and safer walking and cycling environments. In response, the new Master Plan highlighted more connections, better services, and livable and inclusive communities in LTA's future plans (2013, pp. 3–4, 15). Besides expanding existing schemes, such as installing lifts at more overhead bridges and implementing "Green Man Plus" in more pedestrian crossings, new initiatives like "Car-Free or Car-Less Zones" that promote safety and walkability could also directly benefit elders (2013, p. 38–41).

It is notable that, in the past few years, LTA has paid more attention to the objectives of quality, comfort, inclusiveness, and livability, along with the more conventional goals of efficiency and connectivity. With respect to the aging population, more consideration for seniors could be incorporated to make the transport system more inclusive and the environment more livable.

For example, elder pedestrian safety is a multi-faceted issue, which should be dealt with via integrated strategies. Elders are one of the most vulnerable groups of road users, who account for about half of the total pedestrian casualties in traffic accidents in Singapore.[16] Current initiatives for pedestrian safety concentrate on improving existing crossings and regulating walking habits (i.e., increasing time for road crossing, introducing the audio-tactile system at traffic junctions for the visually-handicapped, adding warning signs, and enhancing public awareness of using designated crossings). However, jaywalking is not uncommon among seniors for a variety of reasons, such as their limited capability of walking long distances, their desire to avoid unpleasant walking experiences, and a lack of adequate road crossings that directly connect popular destinations. Optimizing the locations, the number, and the quality of road crossings from a people-centered perspective should be incorporated into the planning for accessibility.

Caring for Seniors: Holistic Affordable Healthcare and Eldercare

Against the present backdrop of increasing challenges from chronic diseases and a pressing need for better personal care and care-giving needs among seniors, the vision of caring for seniors is to create "a seamless continuum of healthcare and eldercare services" through the synchronizing of collaborative efforts by the private sector (made up of companies that can provide healthcare and eldercare services and products), the people sector (healthcare professionals and personnel), and the government sector (through financing, strategic welfare services delivery, manpower development, and setting industry regulation standards on healthcare). This envisioned policy framework focuses on an ecosystem of strong community care services, integrated service planning, diverse care provision, and affordable healthcare and eldercare (CAI, 2006, pp. 33–38).

Prior to CAI, the implementation of the Eldercare Masterplan (2001–2005) had already developed a strong network of community-based eldercare service centers privately run by VWOs, coordinated by the National Council of Social Service (NCSS). This outreach of multi-modal service providers saw the most significant increase in the sector of "day care centers," "home medical services," "home nursing services," and "home help services" from 1998 to 2005. A 2010 study of the integration of services found that a higher number of Senior Activity Centres (SAC) are planned alongside housing estates with a larger population of elder residents and that zones with smaller elder populations were likely to have fewer or no such services

(Chua 2010). These integrated service centers, which are largely sited on the first floor of rental blocks or SAs, help elder residents in both maintaining their health and strengthening their social networks; more SACs are expected to be built in future (MOH, 2012). However, based on the nature of distribution, more SACs could further alienate the hard-to-reach, smaller, and dispersed elder population, simply because it is not a viable provision for the small catchment demographics.

Furthermore, the CAI also recommended adopting family-physicians-based care and one-stop primary care centers to be located in local communities, thereby addressing the service gap for intermediate residential care, fostering integrated day care and rehabilitation centers (CAI, 2006, pp. 36–51). One specific idea of this kind of one-stop primary care center is the Senior Care Center (SCC) (formerly known as Integrated Day Facility). The SCC will provide day care, dementia day care, day rehabilitation services, and basic nursing services. Families can drop off their aged parents at the centers during the day and focus on their work, knowing their parents are being cared for. The SCC would support a network of private family physicians who can more easily provide continuous treatment and care for chronic disease in the elder community rather than on an episodic-care basis. The SCCs will, over time, also offer personal care services at the homes of seniors, including personal hygiene, laundry, and meal services, and serve as staging points for the delivery of other home care services. By 2016, Singapore will have 39 new SCCs to meet the increasing demand for aged care in the community, currently overseen by the Agency for Integrated Care (AIC) (MOH, 2012).

Recent reports support a more distributed healthcare and eldercare model, as centralized hospitals may not be able to provide adequately for the long-term geriatric needs of the elders, especially those requiring further outpatient rehabilitation treatment or those unable to withstand frequent travel to integrated care facilities or to afford inpatient treatment (TODAY, 2013). The newly envisioned arrangement of the healthcare paradigm thus seeks to reduce the strain on hospitals and suggests further intermediate linkages between hospitals and community-based care.

The concept of community-based care further supports the CAI vision for aging-in-place. However, a recent 2011 National Survey highlighted that the overall percentage of elder citizens who went for regular health screenings had dropped from 64.2% in 2005 to its present 50.2%. Also, elder participation in regular health screenings for 2011 registered a further decline in the percentage of participation as compared with 2005, which had seen a general increase of elder participation. The decline through 2011 was apparent in all age groups and both genders (MCYS, 2011). Far more in-depth study, therefore, needs to be done to determine whether there is any relationship between the distribution of healthcare services or mobile screening programs and the incidence of elder participation. In any case, a sustained shift towards a more widely distributed and a more option-based ecosystem of healthcare and eldercare provision is evidently needed in the coming years.

Opportunities for Seniors: Active Lifestyle and Well-Being

The primary objectives of the CAI were to encourage older Singaporeans to "lead healthy and active lifestyles"; to strengthen family ties through intergenerational bonding; to develop social networks; and to contribute actively to their communities through lifelong learning and volunteerism (CAI, 2006, pp. 52–53). The CAI recommended that the government and national agencies would lay the foundations for such goals by committing seed funds for more quality elder activities and programs to promote ground-up initiatives "for seniors and by seniors." The creation of the Golden Opportunities! (GO!) Fund is one example of a program to further this cause (CAI, 2006, p. 56). Another avenue was to raise public awareness through visible channels systematically to inform seniors about activities available to them. Such activities aimed to highlight the importance and benefits of healthy, active living and forming strong social networks to support each other.

To this end, the People's Association (PA) is one such organization that is able significantly to reach out to the senior community. Being within the purview of the Prime Minister's Office, PA oversees the nationwide Community Centres (CCs), which run a range of activities that cater to all age groups in the community, especially seniors. From 2008, PA spearheaded the Wellness Programmes in 12 pilot sites situated within selected CCs in various housing estates. These programs aimed to promote health screening, physical activities, and the formation of interest groups among elders. The overall program is planned gradually to expand its reach to 500,000 seniors over the 87 constituencies island-wide by 2015 (PA, 2010). This Wellness Programme includes the formation of an Active Aging Council within each CC, which would develop unique programs for outreach to its elder residents and work with key stakeholders and grassroots organizations within the community to address active aging issues in areas such as volunteerism, health screening, and recreation activities (PA, 2008). The government's Health Promotion Board (HPB) is also actively involved with the Wellness Programme, providing mobile health screening for chronic diseases and organizing health talks for elders through these CCs as venues.

On another level, PA also runs over 400 Senior Citizens' Clubs (CAI, 2006, p. 59); the National Library Board organizes storytelling programs for children in regional libraries and various housing estates that involve elder volunteers reading to local children. This interrelationship between agency-led initiatives within the urban fabric of Singapore has contributed greatly to the dissemination of programs in various heartlands.

With respect to city and urban infrastructure, the CAI made two proposals. Firstly, that more infrastructural support and facilities be provided to encourage the elder community more actively to participate in sports and other community-centered activities. Secondly, that more public space be made available within housing estates and neighborhood parks for seniors to interact and socialize (CAI, 2006, pp. 57–60). More elder-friendly sports facilities and age-friendly playgrounds equipped with elder-friendly exercise equipment would make it more conducive for an older individual to exercise and engage in sporting activities with his family and within the community, encouraging intergenerational bonding and stronger family ties. The CAI stated that a 2004 survey, conducted by the Singapore Sports Council Mini-Survey on seniors, found that seniors prefer to have activities within their housing estates.

Currently, void decks[17] in public housing estates are often used for more senior-related activities by Senior Citizens' Clubs and Residents' Committees (RC). These void decks are often converted into Senior Citizens' Corners or used as spillover spaces during RC events. Even in the absence of such activities, one can typically observe elders congregating in these spaces to socialize and take advantage of the shelter and seating provided. These observations reinforce the vision that public spaces are integral to promoting and enabling senior communities and for maintaining social networks, developing new friendships, and being more involved in their communities. Often, this also becomes the basis for greater intergenerational cohesion, which is also envisioned by the CAI.

A 1995 National Survey queried about their use of public spaces in and around the housing estates (MCD, 1995, p. 82). However, in later surveys in 2005 and 2011 (MCYS, 2005, 2011), this particular portion of the survey had been omitted and was no longer one of the areas in which their feedback was collected. The 1995 survey seemed to have influenced the way in which the design of urban spaces and community activities were carried out, particularly taking into account the needs of senior citizens. Thus, it is critical to continue seeking feedback from senior citizens in this area to understand whether these initiatives have indeed created the positive outcomes envisioned, and whether these provisions are sufficient for the growing elder population.

Besides spatial planning and the provision of facilities and programs that promote active lifestyles among seniors, there is another aspect of active involvement related to civic participation. In the next section, we discuss the availability of opportunities for civic participation by seniors in Singapore, including volunteering in urban social projects, changing their surrounding environment to enable certain elder-friendly activities, and contributing to the conversations on aging policies and urban development itself.

CIVIC PARTICIPATION

Given Singapore's reputation as a well-established, paternalistic, and hierarchical system of government, it is not surprising to find that the level of civic participation is generally not high. Though many initiatives reviewed in the previous section have been successful, volunteerism, particularly among seniors, is very low (MCYS, 2011, p. 62). However, it is in the government's interest to promote civic participation, so that by taking into account citizens' insights, it "enhances the trust in the legitimacy of and responsibility for political decisions" (Ho, 2000, p. 455). As such, various channels have been established in the past decade for citizens to participate in the development of Singapore's future. For example, in the making of Concept Plan 2001 and Draft Masterplan 2003, public feedback gathered through media, focus group discussions, public exhibitions, and dialogues was incorporated with these major city plans (Yuen, 2005). Such participation was even magnified to an unprecedented scale in the recent "Our Singapore Conversation," a nationwide outreach program to engage citizens in the conversation on Singapore's future, involving more than 47,000 people in dialogues, letters, emails, and social media over the course of one year (OSC Secretariat, 2013). The first dialogue was actually held with 150 seniors in 2012 at Yuhua Village Market and Food Centre, raising issues on the rising costs of medical care and the insufficient healthcare subsidies that contribute to a sense of insecurity. Other topics raised included the need to adjust community facilities better to serve elders, academic stress of their grandchildren, and the rising cost of living (OSC Secretariat, 2013). These collected voices and aspirations "from the ground" are currently being used to review government's policies. Participation is thus introduced in the form of opportunities for civic engagement and providing avenues for involvement in policymaking and even urban physical planning.

Nevertheless, the general sentiment regarding civic participation is often affected by the obstacles presented by obtaining approval by various agencies. Though the CAI cited the example of Kampung Senang Organic Farming in Tampines, which was started in December 1999 through civic volunteers as a volunteer opportunity for seniors (CAI, 2006), these are very rare examples of such ground-up initiatives. Kampung Senang was initiated by resident Joyce Lye, and it took almost seven months to take off after convincing residents in the area and obtaining approval from government agencies for the site of the community garden within the HDB estate.

A recent news report also highlighted the illegal use of state land for farming by a community of residents largely made up of seniors, at the periphery of the Clementi Avenue 4 housing estate. This plot of state land lies on a sloping terrain parallel to a storm-water canal and an old Malaysian railway, recently returned to Singapore ownership. Residents, who previously participated in the unauthorized planting for the last five years, were asked by the authorities to vacate the land within two weeks in March 2012. Resisting the eviction, a retiree and his son collected petitions from their farming community and applied for a three-year temporary occupation license for the contested plot of state land. With the support of their local Citizens' Consultative Committee, a permit was finally obtained.

Users of this community farm, consisting of 15 original seniors and another 15 new residents (priority given to retirees), will now pay an annual maintenance fee of $60 for the use of their individual plots. The farm was officially launched in March 2013 (Channel News Asia, 2013).

The determination of this community of elders not only resisted eviction from this plot of state land but also sought to reclaim their communal activity of farming. On the other hand, it also shows the shift in government's attitude towards finding ways to accommodate new needs of the citizens, yet within an institutionalized framework. Responding to the rising demand for gardening space, especially on the part of senior residents, the Community in Bloom program has been introduced by the National Parks Board (NParks) to offer a more consultative, less prescriptive approach to allow residents to set up community gardens within their neighborhoods, schools, and workplaces. To date, over 600 community gardens have been started, with some 20,000 residents having become gardeners (Leong, 2013).

Evidently, elders want to draw their own boundaries, determine their paths, and craft their own spaces. In a high-density city-state where space is a premium, it can be understood why there are restrictions on the use of land. Civic participation is also often perceived as engagement of citizens within a prescribed framework, such that elders are encouraged to join the programs organized for them, rather than being motivated to initiate programs by themselves. However, with the aging of highly educated and more opinionated baby boomers, it is thus necessary to rethink the strategy in ownership and usage of land, in order to balance the needs between safeguarding space for economic development and providing freedom to personalize space for community development.

MOVING FORWARD

Facing the unprecedented challenges from the magnitude of an aging population but also recognizing opportunities in a high-density urban setting, Singapore's aging policies have been translated into various urban initiatives to support families in providing care for elders, while calling for efforts from all sectors of society to achieve its goal of Successful Aging, including civic participation, private-sector involvement, VWO-run community-based services, and state-led programs. As reviewed in the previous sections, the extent of implementation of each of these urban initiatives can be evaluated based on CAI's recommendations in 2006.

Elder-Friendly Housing

More housing options such as Studio Apartments (SA), one-room rental flats, and 3-Generational (3Gen) flats have been provided by HDB. However, more alternatives could be explored, such as dual-key apartment and retirement housing, which have seen high demand in a few recently launched private developments. In terms of financial programs, several reverse mortgage schemes have been introduced to help elders to monetize their housing assets to meet expenditure in old age. More care, however, should be taken in downsizing programs because relocation of elders may sometimes result in increased disconnection from their familiar urban settings and social networks, contradicting the original goal of aging-in-place.

Barrier-Free Accessibility

Following the revisions in BCA Accessibility Codes to provide safe and unhindered access in homes as well as between buildings and to outdoor public amenities and recreational facilities, HDB has launched a series of neighborhood and housing upgrade programs, including the Housing Improvement Program (HIP), the Lift Upgrading Program (LUP), the Project Lift Improvement and Facilities Enhancement for Elderly (Project LIFE), and Enhancement for Active Seniors (EASE) to allow seniors to age-in-place independently. LTA has also introduced more low-floor wheelchair-accessible buses, while improving barrier-free access in all transport nodes, walkways, and overhead bridges. In addition, the "Green Man Plus" scheme was implemented to allow more crossing time for seniors. As Universal Design becomes an integral part of building design, more coordinated and innovative design of interconnectivity between buildings and public spaces and deeper understanding of elder usage thus become necessary to avoid under- or over-provision of these barrier-free features.

Integrated Healthcare and Eldercare

A network of community-based eldercare service centers, such as Senior Activity Center (SAC) privately run by VWOs, has been introduced since the launch of the Eldercare Masterplan 2001. Family-practitioner-based holistic care for seniors is currently being promoted, with a new one-stop primary care center, that is, the Senior Care Center (SCC), to be built by AIC in matured estates island-wide to provide continuing care for elders and allow them to age-in-place in their communities. Besides ensuring that these healthcare and eldercare infrastructures are adequately distributed, more emphasis should also be placed on their designs, the experience, and the programs within to increase usage of these facilities by elders.

Opportunities for Active Lifestyles

An extensive Wellness Programme was introduced by PA that oversees the Active Aging Council and Senior Citizens' Club within each Community Center, reaching out to its elder residents and addressing active-aging issues such as health screening (in collaboration with HPB), volunteerism, and other recreational activities. The Community in Bloom program was also introduced by NParks to allow residents (often elders) to set up community gardens within their neighborhoods and to engage in gardening and farming activities. Despite the competing land use, more public spaces with better environmental design should be developed for seniors to enjoy outdoor spaces, form social networks, offer spaces for communal living, or initiate new activities to encourage them to lead more active lifestyles.

Moving forward, several areas can be improved to meet the challenge of accommodating the needs of the large cohort of baby boomers who began to retire at age 65 in 2012.

First, more effective coordination and closer collaboration among different government agencies and initiatives are needed to make the entire urban environment more integrated, friendly, and livable for all, without gaps or over-provisions. Inter-ministerial decisions will have to be realized through cross-agency planning to ensure seamless interface among housing units, housing blocks, outdoor spaces, transportation networks, healthcare and eldercare services, and even various events and activities. For example, other than Car-Free Zone, shared street space among pedestrians, cyclists, and motorized vehicles in housing estates could also be further explored, which would require the collaboration among different agencies, such as HDB, LTA, NParks, and Town Councils.

Second, while much has been done for the physiological needs of elders, more understanding about their mental well-being, cognitive function, and social interactions with others in the community is needed to make aging policies and urban development more holistic and effective. Comprehensive research about seniors' preferences in using urban public spaces and in-depth research on baby boomers should be further conducted. In addition, the existing surveys and research on aging issues should also be more closely incorporated into policymaking.

Third, more active civic participation should be encouraged, given that present and future elder citizens are more educated, motivated, and creative residents of the city. A participatory mechanism in the existing sociopolitical environment should be further developed and elevated, not only to enable seniors to join the various programs organized for them, but also to encourage them to initiate their own programs as well as participate in policymaking and the urban planning process to truly realize active aging.

The uniqueness of the high-density city-state of Singapore has closely connected aging policies with its urban development. Due to land scarcity and competing land use, responding to rapid aging has inevitably led to prudent urban planning to ensure sufficient resources for future growth. The contribution of this chapter is the study of linkages between urban development and aging policies, especially in a high-density city-state. A more coordinated, holistic, and participatory approach to urban planning can better cope with the dynamic nature of aging issues and help to achieve the vision of Successful Aging in Singapore.

NOTES

1 In 2011, the total fertility rate was 1.2 (NPTD, 2013, p. 9).
2 Baby Boomers in Singapore are those born between 1947 and 1964 (CAI, 2006, p. i).
3 Singapore has a relatively large non-resident population. According to 2013 data by the Department of Statistics Singapore (DOS), the total population of Singapore in 2013 reached 5,399,200, among which there were 3,844,800 Singaporean residents (including citizens and permanent residents) (DOS, 2013).
4 The 2013 Population White Paper projected that the total population of Singapore would reach 6.5 to 6.9 million, including a resident population of 4.2 to 4.4 million (NPTD, 2013, p. 6).
5 The old-age support ratio is the number of work-age citizens for every elder citizen. The 2013 Population White Paper defined the work-age citizens as those aged 20–64 years and elders as those aged 65 years and above (NPTD, 2013, p. 13).
6 The highest was awareness of nursing homes (36.8%), while knowledge of services distributed within the urban fabric, such as rehabilitative day centers (11.6%) and day care centers for Senile Dementia (7.9%), was far lower. However, the actual utilization of these services in practice tended to be "confined to those who needed them because of their health or social conditions"; and those who were aware of these services but did not use them cited reasons such as not requiring the service or preference to rely on family support (MCD, 1995, p. 61).
7 In 1985, the Ministry of Community Development (MCD) was established from the previously named Ministry of Social Affairs along with cultural portfolios from the previous Ministry of Culture. The new MCD had a mission to promote a robust, culturally vibrant, cohesive, and caring society. In 2000, the MCD was reorganized and renamed as the Ministry of Community Development and Sports (MCDS), with an added mandate and recognition of the importance of sports in national development and community bonding. Then, in 2004, the MCDS was renamed Ministry of Community Development, Youth and Sports (MCYS), with an added signal of the government's commitment to engaging young citizens in national matters. In November 2012, MCYS was then divided into the Ministry of Social and Family Development (MSF) and the Ministry of Culture, Community and Youth (MCCY). See the MSF official website: http://app.msf.gov.sg. In this chapter, the names and abbreviations of these ministries are presented as what they were called at the time when the respective government documents were published.
8 Aging-in-place refers to "enabling people to continue living in the community they are familiar and comfortable with well into old age, amidst family and community support" (IMC, 1999, p. 150).
9 Examples of step-down care are community hospitals, nursing homes, general practitioners doing more work, people taking care of sick people at home, etc. (SG Press Centre, 2009).
10 For example, the "Not-In-My-Backyard" syndrome surfaced from residents affected by the development of eldercare facilities near their residence. The term "Not-In-My-Backyard," or NIMBY, is often used to describe residents who oppose proposals of development projects close to their homes. Such opposition has occurred more and more often in Singapore recently, especially in the heartlands. Recent examples include residents' unhappiness against building an Eldercare Center in a void deck in Woodlands and residents'

resistance to the construction of a nursing home planned by the Ministry of Health (MOH) at Bishan Street, for various reasons such as blockage of views and airflow and worries about congestion. These opposing voices have been heard, but these projects continued to be implemented. For instance, the design of the nursing home in Bishan has been improved by incorporating feedback from the residents, but MOH proceeded to build the nursing home at the originally planned site (The Singapore Law Review, 2012; The Straits Times, 2012).

11 Launched in September 2013, 3Gen flats are designed for multigeneration families to stay under one roof and also support the needs of larger families, which may require an additional bedroom with attached bathroom. Each 3Gen flat will feature four bedrooms and three bathrooms (two en-suites), with an internal floor area of about 115 square meters. Only applicants with multigenerational families, comprising at least a married or courting couple and their parent(s) listed in the application, will be eligible to purchase these 3Gen flats. See HDB InfoWEB: http://www.hdb.gov.sg.

12 Some of the elder-friendly features are non-slip tiles, leveled flooring, grab bars in the bathrooms, pull-cords that linked to the Alert Alarm System (AAS), etc. See HDB InfoWEB: http://www.hdb.gov.sg.

13 The other two strategic thrusts are making public transport a preferred-choice mode and managing road usage (LTA, 2013, pp. 4–9).

14 The Green Man Plus scheme allows senior citizens to tap the senior citizen concession card or the Green Man Plus card on the readers mounted above the standard push button on traffic light poles, which will give, on average, an additional five seconds of green time for pedestrians. Following the successful pilot tests at five locations in 2009, LTA has extended the scheme to 256 crossings as of 2013 (LTA, 2013, p. 40).

15 The trade-off between using scarce land resources for roads or other purposes will become more acute in Singapore. Roads have already taken up 12% of total land area, compared to 14% for housing. Thus, slower growth in the road network is to be expected in the future (LTA, 2013, pp. 14–15).

16 According to 2010, 2011, and 2012 statistics of the Singapore Police Force, about half of those who died in road accidents were aged 60 and older. Of the 29 elder pedestrian fatalities in 2010, 24 died as a result of jaywalking. See www.spf.gov.sg/prints/tp_annual/2012/index_tp_12.htm; www.spf.gov.sg/stats/traf2010_concern.htm. See also Singapore Road Safety Council (SRSC): srsc.org.sg/advisories/children-senior-citizens-road-safety-accidents.

17 Void decks are important common areas for social activities in HDB public housing blocks. Void decks occupy the ground level of the housing blocks, while apartments are usually on the second floor and above. The idea of a void deck was introduced to public housing design in the 1970s by HDB to serve a wide range of uses. Relatively permanent uses include community facilities such as kindergartens, childcare centers, senior activity centers, small shops, etc. Temporary purposes include social functions such as sitting, casual gathering, celebrations, weddings, funeral rites, etc. Void decks also function as "lobby lounges" of high-rise HDB blocks to enhance the sense of belonging to communities (Tan, Loh, Tan, Lau, & Kwok, 1985, pp. 76–77).

REFERENCES

Addae-Dapaah, K. (2008). Age segregation and the quality of life of the elderly people in studio apartments. *Journal of Housing for the Elderly*, 22(1–2), 127–161.

Building and Construction Authority (BCA). (2007). *Code on accessibility in the built environment 2007*. Singapore: BCA.
Building and Construction Authority (BCA). (2013). *Code on accessibility in the built environment 2013*. Singapore: BCA.
The Business Times. (2013, March 31). *HDB to wrap up lift upgrading programme by end of next year*. Singapore: Singapore Press Holdings.
Channel News Asia. (2013, March 3). *Community farm for Clementi residents launched*. Singapore: Channel News Asia.
Chua, C. S. (2010). *Getting the elderly to remain active*. Institute of Policy Studies. Singapore: Lee Kuan Yew School of Public Policy. National University of Singapore.
Committee on Ageing Issues (CAI). (2006). *Report on the ageing population. Ministry of Community Development, Youth and Sports (MCYS)*. Singapore: CAI.
Department of Statistics (DOS). (2013). *Singapore in figures 2013*. Singapore: Department of Statistics, Ministry of Trade and Industry, Republic of Singapore.
Ho, K. L. (2000). Citizen participation and policy making in Singapore. *Asian Survey*, *40*(3), 436–455; University of California Press.
C:\Users\mwaldren\Desktop\Geary QI\6244-0641-Ref Mismatch Report.docx - LStERROR_11Housing Development Board (HDB). (2015). *InfoWEB*. Available from http://www.hdb.gov.sg/
Inter-Ministerial Committee (IMC). (1999). *Inter-Ministerial Committee (IMC) report on the Ageing Population, 1999*. Singapore: IMC.
Land Transport Authority (LTA). (2008). *Land transport masterplan—a people-centred land transport system*. Singapore: LTA.
Land Transport Authority (LTA). (2013). *Land transport master plan 2013*. Singapore: LTA.
Leong, C. C. (2013). *Community in bloom: Giving citizen gardeners room to grow* (Urban Solutions, Issue 3). Singapore: Centre for Liveable Cities. Inter-Ministerial Committee on Sustainable Development.
Ministerial Committee on Ageing (MCA). (2009). *State of the elderly in Singapore*. Singapore: MCA.
Ministerial Committee on Ageing (MCA). (2012). *COS speech by minister Gan on ministerial committee on ageing's initiatives*. Ministry of Health. http://www.moh.gov.sg/content/moh_web/home/pressroom/speeches_d/2012
Ministry of Community Development (MCD). (1995). *National survey of senior citizens in Singapore, 1995*. Singapore: MCD.
Ministry of Community Development and Sports (MCDS). (2001). *Eldercare masterplan (FY2001 to FY2005)*. Singapore: MCDS.
Ministry of Community Development, Youth and Sports (MCYS). (2005). *National survey of senior citizens in Singapore 2005*. Singapore: MCYS.
Ministry of Community Development, Youth and Sports (MCYS). (2011). *National survey of senior citizens in Singapore 2011*. Singapore: MCYS.
Mehta, K. K. (2006). A critical review of Singapore policies aimed at supporting families caring for older members. *Journal of Aging & Social Policy*, *18*(3–4), 43–57.
Mehta, K. K., & Briscoe, C. (2004). National policy approaches to social care for elderly people in the United Kingdom and Singapore. *Journal of Aging & Social Policy*, *16*(1), 89–112.
Ministry of Health (MOH). (1984). *Report of the committee on the problems of the aged*. Singapore: MOH.
Ministry of Health (MOH). (2012, September 28). *More facilities to help seniors age in place*. Available from https://www.moh.gov.sg/content/moh_web/home/pressRoom/pressRoomItemRelease/2012/more-facilities-to-help-seniors-age-in-place.html

Ministry of Social and Family Development (MSF). Available from Official website: http://app.msf.gov.sg/

National Population and Talent Division (NPTD). (2013). *A sustainable population for a dynamic Singapore.* Population White Paper. Singapore.

Our Singapore Conversation Secretariat (OSC Secretariat). (2013). *Reflections of Our Singapore Conversation.* REACH (Reaching Everyone for Active Citizenry @ Home). Government of Singapore.

People's Association (PA). (2008). *People's association wellness programme starter kit.*

People's Association (PA). (2010, May 24). *Nationwide wellness programme for all seniors* [Press Release]. Available from http://www.pa.gov.sg/highlights/nationwide-wellness-programme-151.html

The Singapore Law Review. (2012, October 14). *Not in my backyard.* Singapore: National University of Singapore.

SG Press Centre. (2009). *National day rally speech 2009 Part 2 (Our people).* Government of Singapore. http://www.news.gov.sg/public/sgpc/en/media_releases/agencies/pmo/transcript/T-20090816–1.html

The Straits Times. (2012, September 25). *MOH to proceed with nursing home in Bishan.* Singapore: Sinagpore Press Holdings.

The Sunday Times. (2012, March 25). *Seniors live large in small space.* Singapore: Sinagpore Press Holdings.

The Sunday Times. (2013, June 9). *Schemes to monetise flats "can be simpler."* Singapore: Sinagpore Press Holdings.

Tan, K.J., Loh, C.T., Tan, S.A., Lau, W. C., & Kwok, K. (1985). Physical planning and design. In A.K. Wong & S.H.K. Yeh (Eds.), *Housing a nation: 25 years of public housing in Singapore* (pp. 56–112). Singapore: Maruzen Asia.

TODAY. (2013). Address medical, social needs of ageing population. 22 May. MediaCorp, Singapore.

World Health Organization (WHO). (2007). *Global age-friendly cities: A guide.* Geneva, Switzerland: World Health Organization.

Yuen, B. (2005). Searching for place identity in Singapore. *Habitat International, 29*(2), 197–214.

Canada

5 Lessons Learned from a Canadian Province-Wide Age-Friendly Initiative

The Age-Friendly Manitoba Initiative

Verena H. Menec, Sheila Novek, Dawn Veselyuk, and Jennifer McArthur

INTRODUCTION

The World Health Organization (WHO) launched the Global Age-Friendly Cities project in 2006 (WHO, 2007). Since then, there has been rapidly growing interest in the concept of age-friendliness on the part of policymakers worldwide (Plouffe & Kalache, 2011; WHO, 2010). In Canada, several provinces have introduced age-friendly initiatives (Plouffe et al., 2013; Public Health Agency of Canada, 2010). Manitoba, a midwestern Canadian province, launched its Age-Friendly Manitoba Initiative (AFMI) in 2008 as a way to "support seniors in leading active, socially engaged, independent lives that contribute to healthy aging" (Seniors and Healthy Aging Secretariat, n.d.). To date, 100 communities have joined the Initiative, representing over 80% of the population of the province.

In this study, we discuss lessons learned from the AFMI by drawing on a formative evaluation conducted in 2011. Formative evaluation focuses on an assessment of the process of implementing a program (Dehar, Casswell, & Duignan, 1993; Schoster, Altpeter, Meier, & Callahan, 2012; Stetler et al., 2006). As such, it can provide valuable information on the complexities of implementation, provide an early indication of whether expected outcomes are achieved, and help to shape decisions as implementation moves forward (Stetler et al., 2006).

Our evaluation framework was, in part, guided by the five-phase process proposed in the Age-Friendly Rural/Remote Communities Guide developed in Canada under the sponsorship of the Federal/Provincial/Territorial Ministers Responsible for Seniors (2007): (1) forming a committee; (2) conducting a community assessment; (3) planning projects; (4) implementing projects; and (5) monitoring progress. We also drew on evaluation frameworks proposed in the context of the healthy cities movement (de Leeuw, 2001, 2009; Donchin, Shemesh, Horowitz, & Daoud, 2006; Plümer, Kennedy, & Trojan, 2010). Similar to the age-friendly movement, the WHO

Healthy Cities initiative was designed to enhance community infrastructure in order to promote health (WHO, 1986), with elements of action defined as political commitment to the initiative; establishment of organizational structures to manage change; commitment to developing a shared vision; and development of intersectoral partnerships (Plümer et al., 2010).

In this chapter, we report on five aspects of implementation of the AFMI: organization and management of the age-friendly initiative at the local level; community assessment; intersectoral partnerships; promoting age-friendliness; and projects implemented. We also identify challenges in implementation.

The Age-Friendly Manitoba Initiative (AFMI)

The AFMI is rooted in the WHO conceptualization of age-friendly communities (WHO, 2007), with age-friendliness broadly defined in terms of a range of domains, such as housing, transportation, health care services and community supports, and communication and information. The WHO definition builds on previous work and initiatives on the features that are important to promote independence, health, and quality of life among older adults (Alley, Liebig, Pynoos, Banerjee, & Choi, 2007; Feldman & Oberlink, 2003).

The AFMI is led by the Manitoba Seniors and Healthy Aging Secretariat, the section within the provincial government responsible for issues related to older adults. In five successive intake rounds to date, all 198 municipalities in the province have been invited to become part of the AFMI. Invitations to municipal government are in the form of a letter from the Minister of Healthy Living, Seniors, and Consumer Affairs, jointly signed by the president of the Association of Manitoba Municipalities. All communities that have wanted to join the AFMI were accepted. The initiative is implemented via municipalities as legal entities. For simplicity's sake, we refer to municipalities (urban or rural) as "communities" throughout this chapter.

Communities receive small amounts of funding from the provincial government to help defray some of the costs associated with planning activities or to implement small projects. They are also invited to a one-day orientation workshop led by the Seniors and Healthy Aging Secretariat that provides information on the concept of age-friendliness and identifies ways to get the initiative started (e.g., the importance of forming an age-friendly committee). On several occasions, community connections workshops have also been held in which representatives from participating communities have been brought together to share experiences.

A partnership between the Manitoba Seniors and Healthy Aging Secretariat with university researchers, which was formalized by a grant from a national research funding agency, allowed our research team to work closely with the communities that are part of the AFMI, including conducting the formative evaluation discussed here.

METHODS

Communities

The province of Manitoba has a population of about 1.2 million. About half (55%) of the population is concentrated into one city (Winnipeg), located in the south of the province (Statistics Canada, 2012). The next three largest centers in the province have between 10,000 and 100,000 residents. Consistent with this geographic profile, the AFMI is to a large extent composed of small, rural communities. As all larger centers have joined the initiative, the AFMI includes the majority of the population of Manitoba (81.7%). The present study was conducted between July and September 2011, with 57 communities eligible to participate in the evaluation at that time.

Participants

The goal was to interview at least one individual from each eligible community who could provide an update on the status of the project. This included age-friendly committee chairs, committee members, or knowledgeable individuals working for the municipality. A total of 67 interviews were conducted with representatives of 44 of the 57 eligible communities, for a 77% response rate. In 25 communities, we interviewed one individual. In 19 of the communities we interviewed, two or more individuals provided further insights into the progress made. Interviews were conducted over the telephone and lasted approximately 30 minutes.

Questionnaire

We developed a questionnaire with closed-ended and open-ended questions pertaining to five areas:

- *Organization and management*: Whether an age-friendly committee had been formed, its membership and organization (e.g., whether the committee meets regularly) and the connection between the committee and municipal government (e.g., whether it was endorsed through a formal municipal council resolution).
- *Community assessment*: Whether consultations (most of these were conducted by our research team) were useful and how information resulting from them was used.
- *Intersectoral partnerships*: Whether new partnerships had been formed (and, if so, with whom).
- *Age-friendly promotion*: Whether (and how) the communities had tried to create awareness about the importance of becoming age-friendly. Participants were asked to comment (and provide examples) in four areas: (1) special events or workshops (e.g., for businesses, age-friendly

awareness day); (2) community presentations; (3) development of a website or mention on the community website; (4) articles in the local newspaper or on local radio; or (5) other things, like development of an age-friendly sign or logo.

- *Implementation of age-friendly projects*: Whether specific projects had been implemented. In the case of communities in which we had conducted community consultations, we asked participants to comment on each of the priorities identified at the consultation. For communities in which we had not conducted a consultation, we simply asked participants about the projects they had implemented.

In each of these sections, participants were also asked to comment on the challenges they experienced.

FINDINGS

Findings are presented for each of the five topic areas addressed in the interview. As there was considerable overlap in the challenges identified across the different areas, we discuss these together.

Organization and Management

The majority of communities had formed an age-friendly committee; only three communities had not yet formed a committee and one community's committee was no longer in existence at the time the interview was conducted. The size of the committees ranged from two to 19 members and had diverse representation: elected officials, such as a municipal councilor or mayor (in 83% of communities); older adults who did not represent an organization (80%); senior organizations or senior centers (65%); senior resource coordinators, who are responsible for community services, such as meal programs (60%); town employees (53%); and representatives from recreation departments (48%), health authority (30%), nonprofit organizations (28%), business (23%), clergy (18%), and education (e.g., school principal, 8%).

Most age-friendly committees were quite formally organized in terms of circulating an agenda (80%) and keeping minutes (78%); however, less than half of the committees (45%) met on a regular basis, with many meeting only on an "as needed" basis. In addition, only a third of committees had adopted formal terms of reference, and few (18%) had addressed how they would deal with member turnover.

Although participants from most communities (89%) indicated that the age-friendly committee or the age-friendly project was formally endorsed by municipal council, only three specified that this was done officially through a resolution passed by municipal council. However, in the majority of communities, participants indicated that the age-friendly committee reported to municipal council, typically because elected officials (i.e., councilors, mayor) were members of the committee and therefore relayed information back to council. In five communities, minutes from the age-friendly committee meetings were sent directly to the council.

We also asked whether communities had experienced a turnover in mayor or council, as there had been municipal elections throughout Manitoba in October of 2010. Change in council had occurred in about half (43%) of communities. In a few instances, there was complete turnover in both the mayor and councilors. In some cases, the change was viewed positively (e.g., new council members joined the age-friendly committee), whereas in other communities, the turnover was thought to have a negative impact, as information was not transferred to new councilors, as illustrated by the following comment:

> One of the problems is the election in October. I guess the information's all there but you have to know what you're looking for and what to ask for . . . but there was very little continuity from one council to the next. The transfer of information never happened.

Community Assessment

The majority of communities had a community consultation conducted by our research team (80%) and had received a report summarizing findings. Briefly, the consultations involved two aspects. First, participants completed a short survey to assess their perceptions of the community's age-friendliness and, second, we facilitated a group discussion that allowed participants to identify priorities to be addressed to make the community more age-friendly. Findings from the survey and the group discussion were then summarized in a report provided to the age-friendly committee.

The remaining communities also conducted some type of assessment, with some using a survey developed by our team and others simply inviting residents to bring issues to the age-friendly committee or town council. Community consultations facilitated by our team were perceived as useful, and, in the majority of cases (91%), the consultation report was discussed by the age-friendly committee at one of their meetings. Copies of the consultation report were shared with municipal councils in most communities. Several communities also chose to share the report with various other individuals and groups in the community, including businesses, the senior center, the town office, and, in some cases, all who attended the consultation. One community held a large meeting:

> What we did was offer pizza, and we had about 75 people come out and listen to the report, and we were really pleased. We asked at the meeting for additional people to volunteer for membership on the [age-friendly] committee.

Intersectoral Partnerships

When asked about whether they had developed new collaborations and partnerships since they got involved in the AFMI, participants from about half (48%) of communities indicated that they had done so. Partnerships were established with a range of organizations, including schools, local businesses, senior centers, governmental organizations, and nongovernment agencies. They were developed to build capacity, to raise awareness about the age-friendly initiative, to implement age-friendly projects, and to improve accessibility in local businesses and public spaces. Many participants also indicated that everybody works together anyway and that partnerships did not newly develop because of the AFMI ("It's a small community, and everybody works together").

Age-Friendly Promotion

More than half (61%) of communities had promoted age-friendliness in some way. In about half (55%) of communities, there had been some coverage of age-friendliness in the local newspapers or on the radio, and in 41% of communities, information was posted on local websites. In 45% of communities, an event or workshop was held to promote the concept; the most common event was an age-friendly awareness day. Moreover, 25% of communities developed a logo or sign.

Implementation of Age-Friendly Projects

In examining responses regarding what projects had been implemented, it became evident that participants often did not differentiate between projects or activities that were in existence *prior* to becoming part of the AFMI and those that were implemented *as a result* of joining the initiative. In analyzing the interviews, we therefore took a conservative approach and focused only on those projects or activities most clearly identified as having been implemented after becoming part of the AFMI. Criteria for considering a project or activity as being due to the AFMI were (1) specific mention was made of the involvement of the age-friendly committee in its development or planning; (2) it was identified as being new; (3) its description suggested that it was not part of ongoing activities (e.g., improving sidewalks was frequently mentioned, but this was often described as being part of regular infrastructure maintenance already in place); and (4) the community context suggested that its implementation was plausible (e.g., projects described in the context of one community that had not even formed an age-friendly committee were not included, whereas some major projects in another community that had an extremely active advisory committee and champion were included).

The majority of communities (77%) had implemented at least one project; some communities had tackled two or more projects. The most commonly mentioned projects fell into the areas of outdoor spaces, buildings, activities, and communication.

A number of communities had also enhanced activities for older adults, with many of these focusing on intergenerational activities. As one participant noted:

> We have a teen center that we're on the verge of changing over to a teen and senior center. We've had meetings with the teens about how the seniors would interact with them and how facilities might be improved because the seniors share their space, and we're hoping to use that teen center for the congregate meal program.

One community with a particularly active age-friendly committee and champion was successful in helping to develop new housing for older adults, with more "visitable" housing being planned. Overall, however, communities were less successful in addressing the following issues that they had identified as priorities during the community consultations: increasing and improving housing (e.g., adding new housing units, especially assisted living; upgrading and modifying existing housing units); increasing health care services (e.g., increasing long-term care beds, recruiting more physicians); improving transportation options (e.g., acquiring a handi-van, a transportation option for people with physical disabilities); developing new community facilities (e.g., senior center, wellness center, indoor swimming pool); and acquiring cell phone coverage (in rural areas).

Challenges in Becoming More Age-Friendly

Participants identified a lack of capacity as a major challenge, particularly in smaller communities in which people often felt that they had too many commitments already. As one participant commented:

> The fact that there are just not enough people to go around for all the committees, and it's just one more committee.

A second, overwhelmingly endorsed challenge relates to the funding needed to become more age-friendly:

> Where does the village come up with the money to do the necessary work? Where does the community center come up with the work to do renovations? So it's been a "we have the ideas but we don't have the resources" kind of an issue.

Many participants commented on the need to have both people and funds to make change happen: "Well, the key challenges are always, of course, resources. We need people; we need money."

Lack of direction and leadership was identified by participants as another impediment to becoming more age-friendly. This was expressed in terms of a general lack of leadership on the part of local government: "I think the town needs to make a stronger step toward leading the initiative here." It was also expressed in terms of the functioning of the age-friendly committee: "Another big challenge that we've had is sort of a lack of direction or goals for what we want the committee to do." This lack of direction also affected other aspects, including building new partnerships: "It's sort of not even knowing what projects to start so you can't even really identify partners if you don't know which projects you want to start."

Finding the right person or people to lead the initiative was therefore seen as critical, albeit at times a challenge:

> Lots of people like to talk, they don't like to do the walk. Trying to find those key people in your community that will follow through on getting things . . . going as far as they can to get things done.

Participants further spoke about the difficulties of engaging stakeholders because of their lack of awareness and buy-in to the concept of age-friendliness: "Convincing some of the people that things have to be done." Getting business involved was identified as a particular challenge. As one participant put it:

> Trying to engage the entire community . . . our hardest one is how do we engage the businesses, how do we get them to be aware that we need to be more age-friendly?

DISCUSSION

Given the increasing momentum that the age-friendly movement has been gaining worldwide, generating knowledge about factors that contribute to successful implementation is critical, and literature on the implementation of age-friendly initiatives is starting to emerge (see e.g., Buffel et al., 2014; Keyes et al., 2014). The present formative evaluation provides a first look at the progress that has been made by communities that are part of a Canadian province-wide age-friendly initiative. Given the inclusion of 44 communities, we were able to examine factors that can either facilitate or hinder communities' progress in a wide range of contexts, including urban and rural settings.

Overall, the study documented many positive developments. For instance, all but four of the communities had age-friendly committees at the time the interview was conducted. Most of the committees had a diverse membership and had some connection to local government. The majority of communities had also conducted a community assessment to identify priorities for action.

Moreover, the majority of communities were promoting the notion of age-friendliness in some form (e.g., by holding an age-friendly awareness day), and most had begun to address at least one priority. Most of the projects implemented were small-scale; for example, adding handicapped parking spots, benches, and automatic doors was common. Larger projects were identified as priorities in most communities (e.g., housing for older adults, adding transportation options), but had generally not been addressed, although several communities had started the planning process and one community had success in developing a housing complex for individuals aged 55 and older. That the projects that had been implemented were generally small is not surprising, given that communities had been part of the initiative for only two to four years, and larger projects such as creating affordable housing for older adults could take many years to put into place. In the context of the healthy cities movement, Draper, Curtice, Hopper, and Gorman (1993) proposed a four-to-eight-year time frame to create healthier settings. A similar time frame is likely needed to make communities more age-friendly.

The present study highlights several lessons learned that may assist other jurisdictions in becoming more age-friendly.

Becoming Age-Friendly Requires Strong Leadership at All Levels of Government

Strong leadership is essential at the local level, with the provincial (in a Canadian context) and federal governments playing key leadership roles. At the *local level*, our study shows that communities more successful in implementing projects tended to have a diverse (and sufficiently large), well-organized age-friendly committee (e.g., regular meetings) and support from municipal government (e.g., having a municipal councilor on the age-friendly committee). Conversely, getting the initiative off the ground tended to be difficult in communities that did not have strong age-friendly committees. Difficulties also arose in cases in which the committee was not clear about its role. This speaks to the need for communities to define clearly the role and responsibilities of the age-friendly committee.

Although most communities had support from local government, this support was generally informal. Few communities had council resolutions in place that formally recognized the committee and/or assigned a councilor or municipal employee to sit on the committee. Having such structures and municipal buy-in is essential for sustainability (de Leeuw, 2001; Plümer et al., 2010). For example, about half the communities in our study had changes in municipal council due to elections. In some instances, the turnover meant that information on the age-friendly initiative was not transferred to new councilors or mayors, thereby placing the initiative in jeopardy.

The *provincial government* plays a critical role in providing overall leadership for the AFMI, first, by systematically engaging communities and, second, by continuing to support communities. Thus, combining a top-down (government-driven) approach has been highly effective in the AFMI, reaching over 80% of the population of Manitoba located in 86 communities, with the bottom-up (community-based) approach ensuring that the initiative addresses communities' unique needs. This suggests that, for maximal impact, age-friendly initiatives would ideally be implemented at a larger regional level, rather than at the community level alone. A regional (provincial in the case of Canada) approach is particularly important in engaging rural communities, which may not have the capacity to learn about the age-friendly movement or start to implement it.

The third level of leadership in Canada is provided by the Public Health Agency of Canada, the section within the Canadian *federal government* responsible for the "age-friendly file." Recently, the Public Health Agency of Canada introduced pan-Canadian age-friendly community "milestones" to make explicit expectations for communities participating in age-friendly initiatives (Public Health Agency of Canada, n. d.). Milestones include, for example, establishing an advisory committee that involves older adults and securing a municipal council resolution to support the age-friendly initiative (Public Health Agency of Canada, n.d.). These milestones were adopted by the Province of Manitoba in the summer of 2011. A recognition award is provided to communities that have reached all milestones. The award is recognized by both the Public Health Agency of Canada and the WHO, thus providing national and international recognition as well.

Communities Need Support

Communities, particularly small, rural ones, identified lack of funds and lack of capacity as two major challenges. Having provincial "age-friendly" funding available for communities would therefore be important. In addition, communities could be provided with assistance to identify and access existing funds (e.g., funds available to develop housing projects).

In Manitoba, the partnership among the province's Seniors and Healthy Aging Secretariat and university researchers provided significant support to communities. A staff member was hired to liaise with communities, and the grant provided the capacity to conduct community consultations and provide other support (e.g., meet with key stakeholders to promote age-friendliness, assist with surveys). The vast majority of communities requested help with a consultation, which contributed significantly to moving their age-friendly agenda forward. Thus, developing partnerships with university researchers or nongovernmental organizations is important for the success of the initiative.

Linking the Age-Friendly Community Initiative to Other Initiatives Is Useful

Communities that were more successful in developing action plans or implementing projects tended to link the Age-Friendly Initiative to already existing priorities or strategic plans and initiatives. For instance, there is overlap between the notion of age-friendliness and initiatives focusing on building accessibility, designing "complete streets" to make roadways safer for pedestrians, and creating walking trails for healthy living. Recognizing the overlap between initiatives creates efficiencies in committee structures and planning processes and can facilitate access to funding. Thus, it can help mitigate the two biggest challenges identified by our participants, lack of funds and lack of capacity.

To Be Sustainable, Ongoing Promotion of Age-Friendliness Is Important

The present study shows that communities promoted the concept of age-friendliness in a number of ways, such as holding age-friendly awareness days. Moreover, promotion needs to occur at the provincial level to ensure sustained buy-in at the local level. For example, community support may be difficult if older adults are not considered a priority, as may be the case in communities in which the demographic make-up is such that children and youth are prioritized.

Community Characteristics Are Important

The present findings suggest that, on the one hand, small rural communities tended to experience capacity and funding challenges more acutely than larger centers, but, on the other hand, some made significant progress in implementing age-friendly projects. Moving forward may be easier in rural communities compared to urban ones because of a variety of factors, such as less bureaucracy, an ability to engage residents more easily because "everybody knows each other," and a greater sense of community. For instance, when asked about partnerships, several of our rural participants commented on the fact that everybody works together anyway.

In contrast, larger urban centers tended to be slower in their progress. Indeed, three urban centers had not yet formed an age-friendly committee at the time the present study was conducted. Becoming more age-friendly may be more challenging in urban centers because of the complexities of working across city departments. Creating an intersectoral advisory committee consisting of representatives of various city departments as well as civil society organizations is therefore likely particularly important in urban contexts (Plouffe & Kalache, 2011). Thus, considering the context in which an age-friendly initiative unfolds is important, since the trajectory and timeline of becoming more age-friendly may differ in rural versus urban communities. Rural communities may be able to show progress with small projects quickly but may have greater difficulty tackling larger projects because of funding and capacity challenges and, as a result, have greater need for support. Urban centers may need more time in the early stages (e.g., to engage residents to identify priorities), but may be able to address larger priorities more easily by building on existing infrastructure and processes.

CONCLUSIONS

Our formative evaluation identified a number of challenges that affect communities' progress in becoming age-friendly. Gathering this information in the early stages of an initiative can be used to improve the initiative as it develops (Dehar et al., 1993) and to direct resources toward those communities most in need of support. From a research perspective, as well as from a policy perspective, an important undertaking for the future is to conduct outcome evaluations to determine whether age-friendly projects are being implemented and whether they benefit older adults. Developing age-friendly community indicators is one important task, and work in this area has been initiated in Canada by PHAC and internationally by the WHO. However, indicators are only a first step, as evidenced by the healthy cities movement, given that evaluation of that movement has been quite limited (Donchin et al., 2006; Green & Tsouros, 2008; Plümer et al., 2010; Werna & Harpham, 1995, 1996). The challenge will be not only to use indicators at the local level for communities to monitor their own progress, but also to have the capacity to conduct evaluations across communities to gain a broader understanding of the impact of the age-friendly movement. Such evaluation work will require buy-in from communities, as well as sufficient funds for a coordinating center to allow compilation of data and data analysis. Moreover, it will be important to conduct qualitative studies to examine in-depth whether and how initiatives like the AFMI have affected older residents who live in diverse communities (e.g., rural versus urban communities).

REFERENCES

Alley, D., Liebig, P., Pynoos, J., Banerjee, T., & Choi, I.H. (2007). Creating elder-friendly communities: Preparing for an aging society. *Journal of Gerontological Social Work, 49*, 1–18.

Buffel, T., McGarry, P., Phillipson, C., De Donder L., Dury, S., De Witte, N., . . . Verté, D. (2014). Developing age-friendly cities: Case studies from Brussels and Manchester and implications for policy and practice. *Journal of Aging & Social Policy, 26*(1–2), 52–72.

de Leeuw, E. (2001). Global and local (glocal) health: The WHO healthy cities programme. *Global Change and Human Health*, 2, 34–45.

de Leeuw, E. (2009). Evidence for healthy cities: Reflections on practice, method and theory. *Health Promotion International*, *24*(Suppl. 1), i19–i36.

Dehar, M.-A., Casswell, S., & Duignan, P. (1993). Formative and process evaluation of health promotion and disease prevention programs. *Evaluation Review*, *17*, 204–220.

Donchin, M., Shemesh, A. A., Horowitz, P., & Daoud, N. (2006). Implementation of the Healthy Cities' principles and strategies: An evaluation of the Israel Healthy Cities Network. *Health Promotion International*, *21*, 266–273.

Draper, R., Curtice, L., Hopper, L., & Gorman, M. (1993). *WHO healthy cities project: Review of the first 5 years (1987–1992): A working tool and a reference framework for evaluating the project.* Copenhagen, Denmark: WHO Regional Office for Europe.

Federal/Provincial/Territorial Ministers Responsible for Seniors. (2007). *Age-friendly rural and remote communities: A guide.* Ottawa, Canada: Public Health Agency of Canada, Division of Aging and Seniors.

Feldman, P. H., & Oberlink, M. R. (2003). The AdvantAge initiative: Developing community indicators to promote the health and well-being of older people. *Family and Community Health, 26*(4), 268–274.

Green, G., & Tsouros, A. (2008). *City leadership for health: Summary evaluation of phase IV of the European Healthy Cities Network.* Copenhagen, Denmark: WHO Regional Office for Europe.

Keyes, L., Phillips, D. R., Sterling, E., Manegdeg, T., Kelly, M., Trimble, G., & Mayerik, C. (2014). Transforming the way we live together: A model to move communities from policy to implementation, *Journal of Aging & Social Policy, 26*(1–2), 117–130.

Plouffe, L. A., Garon, S., Brownoff, J., Foucault, M.-L., Lawrence, R., Beaupré, J.-P., & Toews, V. (2013). Advancing age-friendly communities in Canada. *Canadian Review of Social Policy, 68/69*, 24–38.

Plouffe, L. A., & Kalache, A. (2011). Making communities age friendly: State and municipal initiatives in Canada and other countries. *Gaceta Sanitaria*, *25*(Suppl. 2), 131–137.

Plümer, K. D., Kennedy, L., & Trojan, A. (2010). Evaluating the implementation of the WHO Healthy Cities Programme across Germany (1999–2002). *Health Promotion International*, *25*, 342–354.

Public Health Agency of Canada. (2010). *Healthy living e-bulletin May 2010: Theme age-friendly communities.* Available from http://www.phac-aspc.gc.ca/hl-vs-strat/subscribe-eng.php

Public Health Agency of Canada. (n.d.). *How can Canadian communities become more age-friendly?* Available from http://www.phac-aspc.gc.ca/seniors-aines/afc-caa-eng.php

Schoster, B., Altpeter, M., Meier, A., & Callahan, L. F. (2012). Methodological tips for overcoming formative evaluation challenges: The case of the Arthritis Foundation Walk with Ease Program. *Health Promotion Practice*, *13*, 198–203.

Seniors and Healthy Aging Secretariat. (n.d.). *Age-friendly Manitoba initiative.* Available from http://www.gov.mb.ca/shas/agefriendly/

Statistics Canada. (2012). *Winnipeg, Manitoba (Code 4611040) and Manitoba (Code 46)* (table). Census Profile. 2011 Census. Statistics Canada Catalogue no. 98–316-XWE. Ottawa, Canada: Statistics Canada. Available from http://www12.statcan.gc.ca/census-recensement/2011/dp-pd/prof/index.cfm?Lang=E

Stetler, C. B., Legro, M. W., Wallace, C. M., Bowman, C., Guihan, M., Hagedorn, H., . . . Smith, J. (2006). The role of formative evaluation in implementation research and the QUERI experience. *Journal of General Internal Medicine*, *21*, S1–S8.

Werna, E., & Harpham, T. (1995). The evaluation of healthy cities project in developing countries. *Habitat International*, *19*(4), 629–641.

Werna, E., & Harpham, T. (1996). The implementation of the healthy cities project in developing countries: Lessons from Chittagong. *Habitat International*, *20*(2), 221–228.

World Health Organization. (1986). *The Ottawa charter for health promotion*. Available from http://www.who.int/healthpromotion/conferences/previous/ottawa/en/index.html

World Health Organization. (2007). *Global age-friendly cities: A guide*. Geneva, Switzerland: World Health Organization.

World Health Organization. (2010). *Other participating cities announced*. Available from http://www.who.int/mediacentre/news/releases/2010/participating_cities_28_06_2010.pdf

6 Age-Friendly City in Quebec (Canada), or "Alone It Goes Faster, Together It Goes Further"

Suzanne Garon, Mario Paris, Andréanne Laliberté, Anne Veil, and Marie Beaulieu

INTRODUCTION

Over the past few years, the vast majority of public decision-makers and stakeholders recognize the importance of senior citizens remaining in their present homes and communities for as long as possible. Municipalities are paying more and more attention to the development of innovative initiatives to adapt living environments to the needs of aging persons. The World Health Organization (WHO, 2007) has helped to refocus the best contributions from this series of initiatives by publishing *Global Age-Friendly Cities: A Guide* (AFC). This guide explains various actions through eight fields[1] within the competence of cities interested in tailoring their infrastructures and services to older adults' needs.

The Government of Quebec (2012) has recently published a first policy on aging entitled "Aging and Living Together," proposing a community development approach. This approach refers to a different way of acting in the public policy framework and generates social ties (Chaskin, Brown, Venkatesh, & Vidal, 2001). Often discussed but rarely applied in Quebec,[2] this approach promotes the participation of all members of civil society, the municipal apparatus, and elected officials. It enhances collaborative partnership between various stakeholders by sharing information, discussing problems or specific issues, and agreeing on common goals and actions. Prior to this policy are the AFC initiatives in Quebec (AFCs-QC), initiatives also based on the community development approach. Since the implementation and evaluation of seven AFC pilot projects (Garon, Beaulieu, Veil, Paris, & Bigonnesse, 2012; Garon, Paris, Beaulieu, Veil, & Laliberté, 2013), there are today more than 700 municipalities participating in the AFCs in Quebec (Government of Quebec, 2013). In the AFC-QC context, community-building constitutes a critical condition to the overall success and effectiveness of the implementation.

Based on one case study (called here "Maillons" city),[3] this chapter aims to describe the AFC-QC and to explain the collaborative partnership conditions and factors that foster implementation effectiveness. We chose to describe and explain only one case of the AFCs-QC among seven, because of its strong participation and strong program outcomes, especially in light of the coordination and partnerships among stakeholders, which have implemented most of the projects in their action plan and have showed further innovative aspects. In fact, Maillons city offers an illustration of how collaboration can serve as a catalyst for new answers needed to the complex challenges posed by the fast aging population in Quebec.

We first present the methodology behind the evaluation of the AFC-QC. We also describe the AFC-QC design and process through examples and outcomes from Maillons. Finally, from a collaborative partnership perspective, we illustrate several conditions and factors that may determine success of the AFC-QC process.

RESEARCH METHODOLOGY

This case-study approach focused on multiple-cases analyses (Yin, 2009), which is particularly suitable for a realistic evaluation (Pawson & Tilley, 1997). Our data collections are based on mixed methods (Creswell & Clark, 2010) in a short longitudinal design (2008 to 2012).

The data analyzed were collected throughout a series of four focus groups[4] with steering committee members, seven logbooks and press reviews provided by project managers, 25 minutes of the steering committee meetings, the diagnostic and action plan reports, one survey (N = 32) on collaboration (Mattessich, Murray-Close, Monsey, & Wilder Research Center, 2001) and one (N = 32) on organizational networks (Provan, Veazie, Staten, & Teufel-Shone, 2005). In all, data were analyzed by triangulation.[5]

DESIGN AND PROCESS: A THREE-STEP MODEL

The AFC-QC's model involves three steps: the social diagnostic, the development of an action plan, and the implementation of projects. A cycle lasts

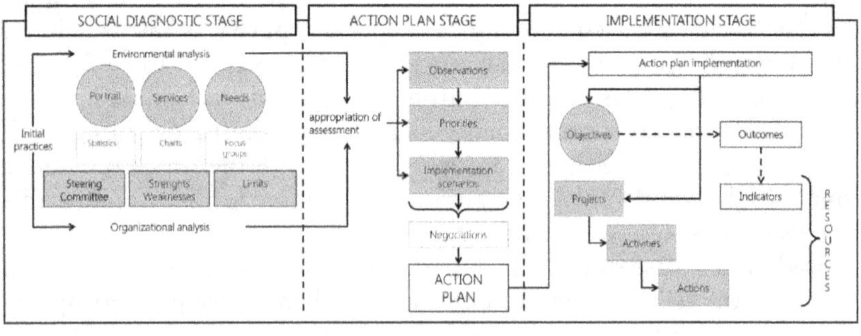

Figure 6.1 Age-Friendly Cities in Quebec Pilot Project Model

over five years, the first 18 months of which carry out the diagnostic and the development of an action plan. Because the AFC-QC's model is more demanding for municipalities than the WHO's guidelines, this action plan must be adopted by the municipal council. The AFC-QC detailed model is illustrated by Figure 6.1.

Steering Committee

None of the steps can be accomplished without the involvement of the steering committee. Its role is to stimulate commitment and to facilitate the achievement of each of the three steps, to implement actions, to disseminate information, and to participate in the mobilization of the community's participants and decision-makers. The steering committee plays an extensive role in facilitating coordination and collaboration. Figure 6.2 shows the composition of the Maillons steering committee in the first and last years of implementation.

First of all, we note that the composition of the steering committee is relatively stable over time. Members have been involved in its deployment according to their initial commitments, depending on the mission of the organizations to which they belong. It should also be noted that the structure of the steering committee is mainly composed of decision-makers at the level of the three municipal councillors, municipal services management, social and health services, or community services. Members all agreed that a structure of collaborative partnership would allow them to arrive quickly at the decision level. In addition, the committee is also balanced with different members from community organizations and associations (usually for or by older adults). In all, more than half of the steering committee members are from the municipal apparatus (6 out of 11, including three municipal councillors). By this strong municipal leadership and its openness to collaborative

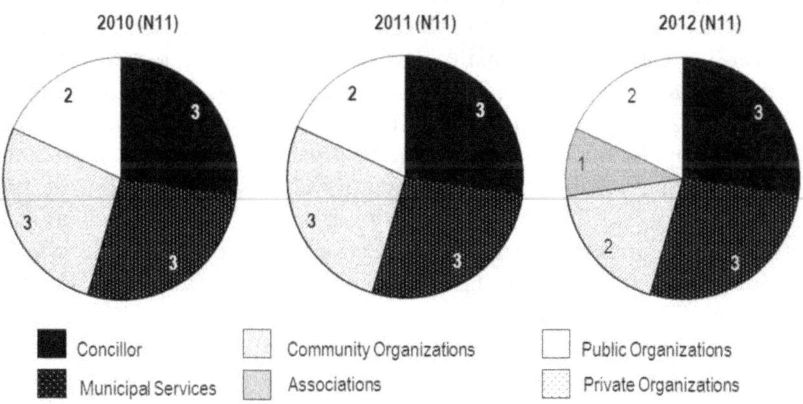

Figure 6.2 Steering Committee: 2010 to 2012

partnership, Maillons developed a strategic positioning of the AFC-QC in the municipal apparatus. Also, it had a coherent decision-making method between the committee and the municipal council and an operating structure linked to other existing collaborative structures.

During the implementation phase, the functioning of the steering committee has been allocated under the responsibility of two structures of existing collaborative partnerships: 1) the municipal committees, including, among others, the Universal Access Policy;[6] 2) the merger of the local seniors' round table, under the influence of the project, with a committee of the health and social services center. Indeed, the work of the AFC action plan soon showed that the duplication of existing committees was not necessary, since the seniors' needs and their organizations were widely discussed at the local seniors' round table. Finally, maintaining the steering committee has allowed, mainly in 2012, the coordination among all members on the various ongoing projects and the planning of future actions.

Step 1—Social Diagnostic

The social diagnostic is essential to the success of the subsequent steps of the AFC-QC process. This operation allows the emergence of a vision shared by all stakeholders on the committee who face the reality of aging in their specific communities. In the case of the Quebec experience, three sources of data collection ensured the rigorousness of the social diagnostic.

The Sociodemographic Portrait

In general, this data collection describes the locality from a population-based statistical perspective, among others from public and municipal organizations or through the Internet. Maillons is a medium-sized francophone municipality located in the vast plains of the St. Lawrence River basin. At the crossroads of an important road network, the municipality takes advantage of a strategic location halfway between the province's two major cities. For over 20 years, the municipality has been experiencing a steady population and economic growth. In 2011, the population was 69,330 inhabitants, or 181.9 inhabitants per square mile (Statistics Canada, 2012). With only 3% of the population, immigration can be regarded as a marginal phenomenon in the municipality. The local economy is largely based on the production of services (79.2%) and, to a lesser extent, on the production of goods (20.1%). In 2011, the median after-tax income of the population was $23,638[7] ($27,675 for men and $19,859 for women) (Statistics Canada, 2012).

In 2011, persons aged 65 years and older accounted for 17.8% of the population. When looking at the population living in the districts of the municipality, it was noticed that some districts have more than 25% of seniors or even 38% in the downtown area. One out of four persons (26.2%) from this group of the population was considered low-income (after tax) according to the 2011 data (Statistics Canada, 2012). These demographic data have had a mobilizing effect for all the stakeholders. The aging of the population was no longer just something theoretical; on the contrary, it was well reflected in the face of this reality!

Grid of Services

This data collection relies on a grid of services, programs, and policies existing in the locality, and gauges services offered and their geographic accessibility. The municipality is served by all the essential services regarding public health and social services, with front-line (hospitals and health and social services centers) and second-line facilities (rehabilitation center, intensive functional rehabilitation unit, day hospital, etc.). There are also several community organizations offering various services, such as household help, support, food assistance, reference, defense of human rights, and recreational and cultural activities. The city has several municipal policies, some of which are in development: family policy, universal access, culture, environment, and recreation and sports. Finally, the municipality has various infrastructures of regular public transport, enhanced by a transport policy adapted for persons with disabilities, as well as a taxi bus service.

Consultation on Older Adults' Needs

This data collection uses focus groups and aims to understand older adults' perspectives of their needs, while taking into account the solutions they offer for a better quality of life and involvement in their community; in accordance with the Vancouver Protocol, eight fields were discussed (WHO, 2007). During the consultation on older adults' needs in 2008, the steering committee established eight focus groups.[8] Results from these consultations were related to issues such as built and social environments.

With respect to the needs of seniors and the built environment, they relate to housing and the living environment, as well as the mobility of seniors. Seniors appreciate their houses, their leased dwellings, and their neighborhoods and want, therefore, to remain there as long as possible. However, some obstacles are forcing many of them to leave their environments: the difficulty of getting certain home-support services, particularly for outdoor maintenance; the perception that the neighborhood is unsafe; the remoteness of services, etc. Although it has several commercial services, the challenge of this mid-sized city is to ensure local services in a context in which shopping centers promote urban sprawl by establishing themselves on the outskirts of the municipality. Furthermore, the ghettoization of the seniors' residences and uncertain safety in the high-rise residential towers in case of fire are two other shortcomings in housing, according to seniors.

Finally, regarding the needs identified in connection with transportation, the difficulties identified by the seniors were access to public transportation, access to parking meters downtown, and the lack of signage for parking spaces reserved for persons with disabilities. As for walking, several seniors complained about the lack of pedestrian lights at many intersections and the very short time to cross streets. Also, the sidewalks and the pedestrian pathways comprise obstacles (e.g., maintenance, snow removal, lighting, difficult relations between pedestrians and people moving in wheelchairs, bicycles, etc.) that severely limit walking distances.

Regarding the identified needs for the social environment, they refer to different dimensions of information, social inclusion, and social participation. First, the major problem identified with regard to the accessibility of health and social services is the lack of information. This situation is exacerbated by the development of new communication technologies such as voice mail, automated responses, and e-mail. In relation to social inclusion, seniors who were consulted observe a generation gap between youth and seniors, which translates into a lack of courtesy or respect. In addition, the various obstacles identified with regard to the social participation of seniors concern the unavailability of public premises for seniors' voluntary activities, succession issues in clubs and associations for seniors, and the scarcity of low-cost activities. Physical activities are not integrated into daily life for several of them (caregivers and sick people) and the range of activities for seniors is uneven depending on the neighborhoods. Opportunities to become a volunteer are not well known. Information gaps on services and activities offered may also limit the participation of seniors.

In general, the seniors interviewed appreciate the quality and diversity of services available to them. However, they have identified many needs that are still inadequately met.

Step 2—Action Plan

After having taken into account older adults' needs and having appropriated the appraisal of the community, the steering committee draws up an overview of the situation. First, the steering committee highlights the overall findings of the social diagnostic (strengths and weaknesses) and defines values and policies that will guide the municipality and its partners in their interventions to improve older adults' lives.

Based on findings, the steering committee members prioritize actions. Then, a plausible and efficient implementation scenario emerges in terms of available resources. The recommended action plan model is based on a logic model approach (Funnell & Rogers, 2011) resting on the principles of results-based management (CIDA, 1999).

The steering committee of Maillons gave priority to 24 actions to implement in order to improve the seniors' quality of life and to encourage their participation in the life of the community (Allie, 2011). The planned actions include a variety of fields, ranging from respect and inclusion of seniors to transportation and pedestrian traffic. Activities have also been planned in the field of information. For example, the activity called "Age-Friendly Firefighters" informs seniors about fire prevention.

The action plan requires the solicitation of partners associated with the AFC's eight fields of action because it contains projects aimed at the built and social environments. The case has demonstrated a broad scope of actions at the municipal level (police, fire, public works, urban planning, library), as well as at the level of civil society and public agencies (social and health services center) and community organizations (round table, seniors' clubs, public education associations, retirees associations, etc.). However, each has not been continuously sought during the AFC-QC process, but they all collaborated in their own way, in their mission, or in their willingness to commit themselves.

Moreover, projects contained in the action plan demonstrate that the steering committee has not viewed the seniors as a homogeneous group. The projects target seniors as a whole, both men and women, as well as those with moderate and low income. Several actions also affect the loss of people's mobility (seniors and persons with disabilities) and are directly related to the Universal Access Policy. The diversity of stakeholders involved in the action plan and the nature of the planned activities demonstrate the willingness of the steering committee to work in an intersectoral manner to meet the needs of the greatest number of seniors, thus responding to their diversity.

Step 3—Implementation

The last step is to plan and organize the necessary resources for the implementation of actions and to carry out a follow up. In fact, the implementation must produce the expected results of the action plan and contribute to achieving the objectives set by the steering committee. Some committees engage quickly in actions they consider timely or for which opportunities arise. Sometimes, projects are somewhat more important and require additional fundraising. The funding provided by the Quebec Government to the AFC-QC mainly pays the project manager's salary.[9] Part of the funding was allocated for some of the projects, but others required substantial human and material resources. To offset these funding issues, although there is no requirement for steering committees to do so, some have responded proactively by quickly seeking new grants.

Examples of Projects

In 2010, the steering committee organized a major campaign that promoted a positive image of seniors. The objective of this project was to change the perceptions that can sometimes be tinged with ageism (by both young people and elders); the campaign, therefore, builds on positive aspects: the importance of maintaining cordial and respectful relations and the enhancement of the role played by seniors in our society, often on a volunteer basis. For the occasion, banners, posters, and advertisements were disseminated throughout the city to highlight the contribution of seniors in the community. The very large part of this promotional material was created from real-life settings of older adults in their milieu. These people took part in photo shoot sessions and agreed to see their pictures widely disseminated, as on municipal buses or in public places (such as municipal offices or the public library).

Another example regarding social environment consisted of making the International Day of Older Persons on October 1st a special time for seniors' organizations, seniors' groups, the health and social services center, and the municipality jointly to organize a day offering several opportunities to inform, through conferences and information booths, on subjects of great interest to them. Year after year, this meeting has become more and more essential for all stakeholders who take an interest in seniors. The organization of this intersectoral event allows stakeholders to reach out to seniors of different stripes (recreation, training, services to seniors, etc.), and implicitly promotes social inclusion of seniors, while ensuring a wide dissemination of the invitation to participate.

The steering committee has also worked on exercise routes as part of a project for seniors called "Active Tuesdays." The objective of this project is to promote physical activity and to give seniors opportunities to create social ties. Several stakeholders were involved in the organization of this sporting event: the municipal services through the Parks and Green Spaces Committee and the family policy, with community organizations, retiree associations, and a dozen seniors' clubs, for an overall total of 22 groups, and those responsible for activities in seniors' residences. This sporting event has also led to the introduction of a training unit in one of the municipality's parks.

The year 2012 did not generate new development, since the first four years of the project as a whole were combined as stipulated in the original agreement with the government, which encouraged sustainable and consolidated projects.

CONDITIONS AND FACTORS OF SUCCESS

To illustrate collaborative partnership conditions and factors that may determine success of the AFC-QC process, we divide this section according to the AFC-QC three-step model.

Steering Committee: A Rooted Network

The creation of a core group of individuals rooted in the community—mainly stakeholders from the municipal apparatus, political representation, and public and community organizations—represents a critical condition of the AFC-QC's process. An example in Maillons of this unique condition is the establishment of an adapted physical training unit for older adults. This project is commonly found in the AFC-QC outcomes. However, the implementation effectiveness relies on the process rather than the outcome. Indeed, in Maillons, the whole implementation process of this project was based on a collaborative partnership: 1) older adults have identified the need for an adapted physical training unit in a public park during the consultations; 2) they have also chosen the type of unit and its location in the city; 3) the municipality has proceeded to its safe installation; 4) the local health and social services center (in partnership with the municipality) has facilitated activities with a kinesiologist; and 5) the municipality publicized the park and the activity in senior homes and clubs. In short, the process was based on a collaborative effort of various stakeholders such as municipal services (civil engineering and recreation), private (senior homes), and public organizations (health and social services center), as well as community organizations and associations (senior clubs).

Social Diagnostic: Towards a Common Goal

From the consultation process with older adults, we have observed the following: The way we proceed, according to the needs identified by older adults themselves, changes the traditional top-down approach in which service providers determine needs and how to respond to them.

In fact, through the consultation process, older adults have introduced ideas that might not have been introduced by service providers. For instance, with the social diagnostic, steering committee members in Maillons acknowledged that a portion of the older adult population was living alone and left behind by service providers. The appropriation of the diagnostic based on consultations allows committee members of Maillons to acknowledge older adults' needs and to challenge the improvement of the living conditions of older adults—a challenge that cannot be undertaken single-handedly.

Action Plan: A Commitment

In Maillons, the action plan (see Box 6.1) was fully integrated in the agenda of municipal services (police and fire departments, public works, urban planning, recreation, and community life), in the priorities of community organizations and associations, and in the practice of health and social services centers, and it constitutes a condition of implementation success.

BOX 6.1 ACTION PLAN'S SAMPLE—RESPONSIBILITIES, PARTNERS, AND REMARKS SECTION

Physical and Leisure Activities

Very wide, small steps, beginning exploratory work: pooling of activities offered, expectations of seniors and challenges of organizations, means to disseminate information . . .

Motivation of groups to work together

In collaboration, the community centers and dozens of seniors clubs, retiree associations, men's and women's associations, persons responsible for activities in seniors' residences, etc.

Link with the Parks and Green Spaces Committee (master plan)

Family Policy: parks, infrastructure adapted to the needs of different groups

Raising awareness among elected officials and those responsible for new projects and infrastructure.

To be explored: Promoting walking as physical activity, but also as an opportunity to create social ties: meeting places (e.g.: Active Tuesdays), suggested itineraries (e.g.: from one park to another), exercise routes, rest areas with benches and other facilities.

All committee members acknowledged each other's missions and interests and clarified the responsibilities attached to the action plan in relation to their combined expertise. In the municipal stakeholder's words: "The concern of the municipality regarding the action plan is not to end up with all the actions to implement, but to foster partnerships and that all stakeholders do their fair share, due to their role, as well as to avoid taking the place of others" (Focus Group, 2012).

Implementation: Acting Together

The Power of Networking

Networks of collaborating municipal apparatus, as well as public and community organizations, give an understanding of the structure of relationships among steering committee members. Figure 6.3 indicates that this case was particularly well networked from the outset; that is, members already belonged to several organizations in addition to those where they were formally working. Results of the 2010 networking questionnaire allowed us to identify 51 links of belonging shared among 11 respondents. The majority of these links were related to community organizations (28 links), as well as to the municipality (15 links). The numbers of listed organizations, as well as their diversity, allow us to see a significant extended network of all the players related to the AFC committee. In 2011, 41 links of belonging are in the municipal (n = 16) and community categories (n = 19), while 45 links of solicitation[10] are mostly directed to community organizations (53%) and show an interesting proportion (22%) facing the municipal sector. These results can be explained by the fact that the community organizations wished to play important roles in improving social environments. The people's network with disabilities is also very present, given that the elements of universal accessibility are important for seniors too. In 2012, Maillons decreased its activities while its committee had concerns about possible ways to finance projects that appear viable and had strong values for stakeholders. Thus, of the 45 links of solicitation shared in 2011, there were only 23 links in 2012, most of them still attached to the community sector (52%).

As implementation progressed, it became interesting to consider the organizations solicited by committee members in their efforts to promote the success of the AFC-QC projects. It can easily be seen from Figure 6.3 that the relationships between municipal services and community organizations intensified in 2011. However, it is also clear that these solicitation links are outside the formal sphere of the organizations to which the members belong. This observation indicates that relationships are well embedded during the AFC-QC's implementation. Also, the solicitation links networking reflects the efforts made by committee members to deploy the AFC-QC's action plan. The network data reveal a vast configuration of membership links and connections that reflects this committee's dynamism about collaborative partnership. As claimed by a member of the steering committee: "The AFC-QC allows us to [. . .] work in collaboration partnership [. . .]. We cannot work in silos anymore" (Focus Group, 2012). The members have invested their networks, relied on the AFC-QC, which helped formalized mechanisms (e.g., task forces, sectorial subcommittees, etc.) and strengthened its action. In the end, we note the central role played by the

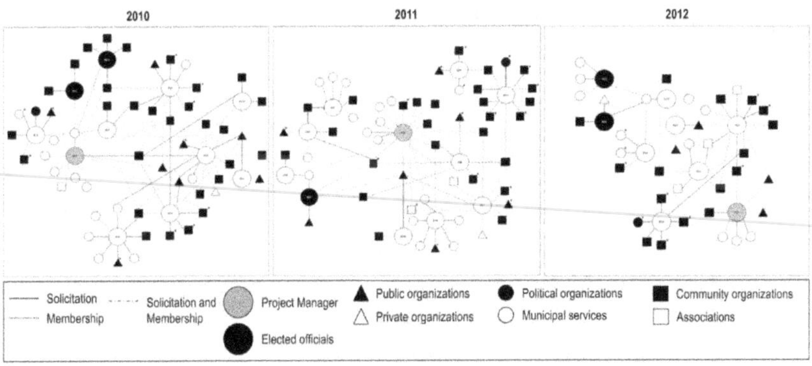

Figure 6.3 Maillons—Results on the Organizational Membership and Solicitation Networking in 2010 to 2012

project manager, which involves several organizations and related instances at the municipal and community levels, as well as health and social services.

The Importance of Collaboration

The transition between the action plan and its execution was analyzed in terms of cumulative effects on the collaboration between partners. The steering committee of Maillons has carried out the action plan fully by ensuring a good collaboration among various stakeholders. This collaboration partnership can be considered an outcome per se: "Well, if we go back five years ago, our organization was working all by itself. Now, all organizations are here and when it's time to collaborate they all do: that's what makes the success" (Focus Group, 2012). Indeed, having a common goal and perceiving positive impacts after the implementation of projects can ensure collaborative partnership. Figure 6.4 shows that, in 2011, all the dimensions examined have increased, but the "aim" dimension particularly stands out from the others.

Over the past three years, the last year has seen a slight decline on most of the dimensions, except for the environmental dimension, which is maintained. The only dimension that stood out was that of resources, which were deemed inadequate. More specifically, in 2012, the municipality brought more attention to different universal accessibility projects with regard to cities' built environment responsibilities, which required less input from the other stakeholders. Also, to address the problem of financial sustainability to support implemented projects and to develop new projects, Maillons's steering committee members mobilized themselves and their networking relations to apply for government grants. In other words, this case, which had a heterogeneous collaborative partnership, accomplished much more

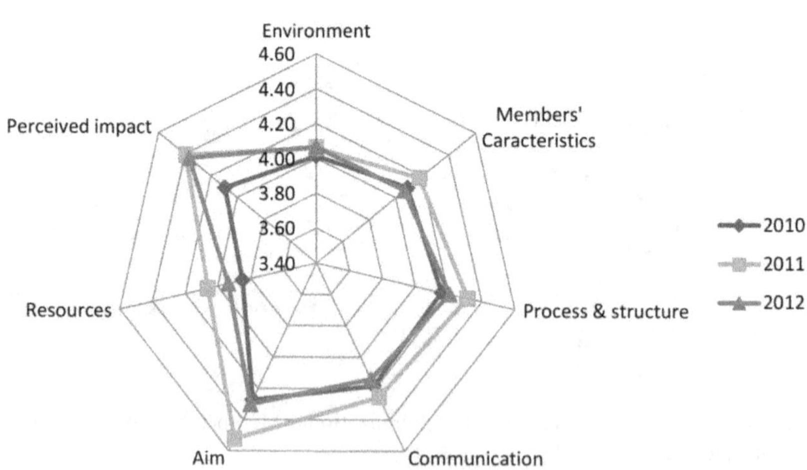

Figure 6.4 Maillons—Results of Collaboration in 2010, 2011, and 2012

through its age-friendly city initiative. This demonstrates the importance of the strength of the partnership relationship as a critical difference in what was or can be accomplished.

CONCLUSION

This chapter described the AFCs model in Quebec (AFC-QC) and explained, through a specific case study, the collaborative partnership conditions and factors that foster implementation effectiveness (see Table 6.1). In doing so, we saw several conditions and factors that may determine the success of the AFC-QC process.

Although with the favorable context prevailing in Quebec (dynamic Seniors' Secretariat, subsidies, social discourse for active aging, fighting structural ageism, government policy on aging, program to counter mistreatment of older adults, etc.), we acknowledge that all of these conditions and factors have contributed to the success of the AFC-QC.

In the wake of the AFC-QC model, the importance of coordinating efforts and fostering collaborative partnerships among stakeholders is clear. Indeed, despite the participation of older adult organizations and associations in the process, they encounter important limits in the capacity of implemented age-friendly environments. The AFC initiatives should lead to a transformation of the traditional top-down approach, not by promoting only a bottom-up approach, but by a collaborative partnership or, in Paquet's and Wilson's (2011) words, a "small governance approach." Instead

Table 6.1 Main Collaborative Partnership Conditions and Factors in the AFC-QC

Steering Committee
• The steering committee members should be rooted in the community and come from diverse sectors. • An elected official and an administrative officer should sit on the committee. • A project manager should be dedicated to the AFC-QC's process.
Diagnostic and Action Plan
• The public consultations with older adults are essential to strengthen involvement of stakeholders around a common goal during diagnostic. • The action plan should commit various stakeholders to ensuring implementation.
Implementation
• The stakeholders should use their networks and share responsibilities during implementation. • A committed elected official is essential to orient political priorities and, therefore, resources (financial, human, material, etc.) available for implementation.

of maintaining a hierarchical, centralized, and authoritarian approach, all stakeholders should change to a pluralistic, participatory, and collaborative approach. This change in attitudes and practices "[. . .] appears to be better equipped to cope with these polycentric coordination challenges, and to reconcile the variety of belief systems at play into a workable accommodation that takes full advantage of the dispersed information, resources, and power under the control of the different stakeholders" (Paquet & Wilson, 2011, p. 2).

However, beyond the necessity to change and communicate with the population, it is of the utmost importance to have a simple model on which there is consensus among stakeholders (Minkler & Wallerstein, 2005). Where the AFC-QC process was evaluated successfully, such as in Maillons, the community-building model was familiar to the municipal administration and the other stakeholders.

To conclude, the importance of collaborative partnership in the AFC-QC leads us to a famous community-building quote: "Alone it goes faster, together it goes further." With that in mind, we get closer to the ideal of a "society for all ages," as enacted in 1999 by the United Nations (United Nations, 1999).

NOTES

1 Transportation, housing, social participation, respect and social inclusion, civic participation and employment, communication and information, community support and health services, outdoor spaces and buildings.
2 The municipal authorities in cities generally establish their own policy and will subsequently validate the social acceptability by formal public consultation. Policies are rarely coupled with an action plan, and often the concrete measures are controlled by the municipal agenda.
3 "Maillons" is a fictive name to protect the identity of the actual community. We are unfortunately unable to provide its name because we are bound by the duty of confidentiality according to regulation RGM-CA-59D, which binds our research center to the ethics committee, itself acting under article 21 of the Quebec's Civil Code.
4 The data collections for this case have included four focus groups with the steering committee in order to understand the stakeholders' perspective on older adults' needs.
5 For more information about this model, the evaluation process, and the methodological approach used in the AFC-QC, refer to Garon and colleagues (2012, 2013).
6 *Universal Accessibility* Policy (in French: *politique d'accessibilité universelle*) attempts to ensure equal access to services for all, irrespective of any disability matter, ensuring accessibility of public buildings and public spaces, as well as the accessibility of disabled and older people in their own homes.
7 All amounts are in Canadian dollars. In 2011, the average exchange rate between Canadian and US dollars was 0.98 CAD or 1.01 USD.
8 1) Seniors from 65 to74 years old with low socioeconomic status; 2) seniors from 65 to 74 years old with average socioeconomic status; 3) seniors aged 75 and older with low socioeconomic status; 4) seniors aged 75 and older

with average socioeconomic status; 5) caregivers of an older adult; 6) public services; 7) private services; 8) community-based and voluntary services.

9 In Maillons, the project manager has been integrated in the municipal apparatus with the creation of an officer position of community policies and programs.

10 Link of solicitation: seeking support from organizations for AFC project.

REFERENCES

Allie, S. (2011). L'approche transversale: au service des aînés et de l'ensemble de la population [The cross-cutting approach: serving the seniors and the population as a whole]. *Vie et vieillissement (9) (Life and Aging Journal) (9)*, 1, 48–53.

Chaskin, R., Brown, P., Venkatesh, S., & Vidal, A. (2001). *Building community capacity*. New York: Aldine de Gruyter.

CIDA. (1999). *Results-based management in CIDA: An introductory guide to the concepts and principles*. Ottawa: Canadian International Development Agency.

Creswell, J.W., & Clark, V.L.P. (2010). *Designing and conducting mixed methods research*. Thousand Oaks, CA: Sage.

Garon, S., Beaulieu, M., Veil, A., Paris, M., & Bigonnesse, C. (2012). L'expérience québécoise du programme « Villes-Amies des Aînés » de l'OMS : l'implantation dans sept projets pilotes. In J.-P. Viriot-Durandal, C. Pihet, & P.-M. Chapon (Eds.), *Les Défis Territoriaux Face Au Vieillissement* (pp. 69–88). Paris: Documentation française.

Garon, S., Paris, M., Beaulieu, M., Veil, A., & Laliberté, A. (2014). Collaborative partnership in age-friendly cities: Two case studies from Quebec, Canada. *Journal of Aging and Social Policy*, 26, 73–87.

Funnell, S., & Rogers, P.J. (2011). *Purposeful program theory: Effective use of theories of change and logics models*. San Francisco: Jossey-Bass.

Government of Quebec. (2012). *Aging and living together: At home, in one's community, in Québec*. Ministère de la Famille et des Aînés and Ministère de la Santé et des Services sociaux, Quebec. Available from http://aines.gouv.qc.ca/documents/politique-vieillir-et-vivre-ensemble_English_version.pdf

Mattessich, P.W., Murray-Close, M., Monsey, B.R., & Wilder Research Center. (2001). *Collaboration: What makes it work*. Saint Paul: Amherst H. Wilder Foundation.

Minkler, M., & Wallerstein, N. (Eds.). (2005). *Community organizing and community building for health*. New Jersey: Rutgers University Press.

Paquet, G., & Wilson, C. (2011). Co-gouvernance collaborative en tant que système de prospection. *Optimum*, *41*(2). Available from http://www.optimumonline.ca/article.phtml?lang=french&id=386http://www.optimumonline.ca/article.phtml?lang=french&id=386

Pawson, R., & Tilley, N. (1997). *Realistic evaluation*. Thousand Oaks: Sage.

Provan, K.G., Veazie, M.A., Staten, L.K., & Teufel-Shone, N.I. (2005). The use of network analysis to strengthen community partnerships. *Public Administration Review*, *65*(5), 603–613.

Statistics Canada. 2012. *Focus on "Geography Series," 2011 Census* (Product no 98–310-XWF2011004). Ottawa, Ontario: Government of Canada.

United Nations. (1999). *A society for all ages—through the international year of older persons*. Available from http://www.un.org/en/globalissues/ageing/

World Health Organization (WHO). (2007). *Age-friendly cities: A guide*. Geneva, Switzerland: World Health Organization.

Yin, R.K. (2009). *Case study research: Design and methods*. Thousand Oaks, CA: Sage.

United States

7 Portland, Oregon

A Case Study of Efforts to Become More Age-Friendly

Alan DeLaTorre and Margaret B. Neal

INTRODUCTION

In this chapter, we detail the efforts of Portland, Oregon, to better serve an increasingly older population and to better utilize the skills, talents, and resources of older adults. We describe the Age-Friendly Portland initiative, including the context, conditions preceding the effort, previous relevant work, structure of the initiative, actions taken to date and planned, and accomplishments and outcomes. We conclude with a discussion of constraints and implications.

CONDITIONS AND FACTORS PRECEDING THE AGE-FRIENDLY PORTLAND INITIATIVE

A number of factors have influenced the age-friendly work undertaken to date in Portland, including the city's/region's demographic characteristics, geography, infrastructure, economic and political environments, the services available, cultural characteristics, and previous research and planning efforts.

Demographic Characteristics

Portland, the county seat of Multnomah County, is the largest city in Oregon; in 2012 it had a population of 603,106, with 10.4% aged 65 or older (U.S. Census Bureau, 2014). Portland also is the largest city in the greater bi-state (including Washington) area, with the Portland-Vancouver-Hillsboro Metropolitan Statistical Area being the twenty-fourth largest in the United States in 2012 (U.S. Census Bureau, 2013). This region is projected to increase from 2,265,500 in 2010 to 3,050,100 in 2030, a growth of 34.6% over those two decades (Metro, 2009). Table 7.1 shows changes in the 65 and older population in the United States, in Oregon, and in the Portland region. Especially noteworthy is the projection that the 65 and older population will more than double during this time frame, from approximately

Table 7.1 Overview of the Number, Proportion, and Growth Rates of People Aged 65 and Older in the United States, Oregon, and Portland-Vancouver Metropolitan Statistical Area (MSA) from 2010–2030

Category	United States	Oregon	Portland-Vancouver MSA
Number of persons aged 65+ in 2010	40,267,984	533,533	251,000
Number of persons aged 65+ in 2030	72,092,000	881,957	518,100
Growth rate in number of persons aged 65+ from 2010 to 2030	79.0%	65.3%	106.4%
Proportion of persons aged 65+ in 2010	13.0%	13.9%	11.1%
Proportion of persons aged 65+ in 2030	19.3%	18.2%	17.0%
Growth rate in proportion of persons aged 65+ from 2010 to 2030	48.0%	31.0%	53.3%

Sources: United States: 2010 estimates are from *The Statistical Abstract of the United States* (U.S. Census Bureau, 2012) and 2030 projections are from *2008 National Population Projections* (U.S. Census Bureau, 2008); State of Oregon: 2010 estimates are from *The Statistical Abstract of the United States* (U.S. Census Bureau, 2012) and 2030 projections are from *Interim State Population Projections* (U.S. Census Bureau, 2005); and Portland-Vancouver MSA: all forecasts are from *20 and 50 Year Regional Population and Employment Range Forecasts* (Metro, 2009).

251,000 in 2010 (11.1% of the region's population) to 518,100 in 2030 (17.0%), representing a growth rate of 106.4% in the number of people aged 65 and older and an increase of 53.5% in the proportion aged 65 and older. Each of these figures is greater than those for the state of Oregon and the United States.

Portland's regional population is relatively homogenous with respect to race and ethnicity. According to American Community Survey data reported by Hammond (2009), the greater Portland region was the fifth whitest in the United States in 2007; 78% of the region reported to be non-Hispanic whites and whites who did not claim an additional identity. Portland's racial and ethnic diversity has increased recently, however, with the non-white population growing from 15% of the city's population in 1980 to 27% in 2010 (City of Portland, 2012a). Another demographic trend is the region's ability to attract and retain college-educated migrants under age 40 at some of the highest migration rates among the 50 largest U.S. regions (Jurjevich & Schrock, 2012). This same study also found that the region outpaced those regions in attracting and retaining empty-nester and retired college-educated migrants aged 40 and older.

Geography, Transportation Infrastructure, and Housing Stock

Portland is located in the upper northwestern area of Oregon, bordered on the north by the Columbia River (separating the city from the state of Washington), dissected by the Willamette River and bordered on the east and west by mountain ranges. The city's west side is hilly, while the east side is relatively flat. The city has been organized into five areas—North, Northeast, Southeast, Southwest, and Northwest Portland. The Willamette River divides the east and west sides, and Burnside St. separates the north and south sides. Downtown Portland, which is on the city's west side, houses a large number of both upper- and lower-income residents, with both high-end condominiums and apartments and public and subsidized housing, including single-room occupancy hotels. According to the City of Portland (2009), nearly 35% of Portland's housing units were built before 1940; another 25% were built from 1940 to 1960, and housing maintenance concerns exist. The aforementioned report from the city also detailed that in 2007, 61% of the total housing units were single-family, detached-style homes, although permit data suggested that multi-family units grew as a proportion of the housing stock beginning in the late 1990s.

A report prepared for Portland's regional government (Neal et al., 2006) detailed several age-related trends in housing and transportation demand in the region: those aged 55 and older were most likely to own their homes; white, older adult householders were the most likely to own their homes; women were less likely to be homeowners than men. Renters had lower incomes than owners; more than 60% of those aged 55 and older lived in single-family, detached housing; there was an increase from 1995 to 2002 among baby boomers and older adult households in the number who lived in a neighborhood with higher density apartment buildings (it was not clear whether that was due to increasing density in the region or preference for higher-density living); and Portland ranked highly, compared to other similarly-sized regions in the United States, for severe housing cost burden, which was seen to limit housing options, especially for renters.

With respect to transportation, Portland has frequently ranked among the top U.S. cities for public transportation and pedestrian friendliness (TriMet, 2014). Inequality in options exists within the city and the region, however. For example, East Portland, home to nearly 165,000 people, can be particularly hard to navigate on foot, by bicycle, or on public transit (City of Portland, 2012b), and, with respect to older adults, the Milken Institute (Chatterjee & DeVol, 2012) ranked the Portland region only 25th among 100 large metro areas in transportation for older adults, with poor marks for the average fare paid by those aged 65 and older (81st of 100) and for public investment in transportation for older adults (86th of 100). At the same time, the region ranked highly with respect to number of passenger trips (10th of 100) and the percentage of people living within proximity to a grocery store (13th of 100). The City of Portland (n.d.) and Metro (2014) have both started active transportation programs that aim to create a transportation network that will enable the majority of citizens to walk, bike, and ride transit to meet their daily needs. These efforts are important to older adults because they increase the accessibility of pedestrian environments (e.g., curb ramps, street crossings, street furniture), improve air quality and physical activity, and may reduce mobility costs.

Economic Resources

According to the Portland Business Alliance (2010), the region has been facing an economic crisis due to job loss and stagnant-to-declining wages. The report predicted that future revenues would fall short of the level needed to cover expenditures and that, over the next 10 years, additional service cuts would be needed to balance budgets. The region's average wage, measured as a percent of the national average for metro regions, declined from its peak of about 102% in 2000 to 96% in 2008; the region's per capita income began lagging behind the U.S. average in the mid-2000s and, in 2008, was $39,000 compared to the U.S. average of $41,000 (Greater Portland Pulse, 2011).

Within Portland, economic inequality is a concern, with fewer Portlanders sharing in economic gains in the region (City of Portland, 2012a). Research from the Coalition of Communities of Color (Curry-Stevens, Cross-Hemmer, & Coalition of Communities of Color, 2010) found that in 2008, communities of color earned half the incomes of whites ($16,636 compared to $33,095 per year); poverty levels were more than double in communities of color compared to whites (33.3% and 12.5%, respectively); and communities of color had unemployment rates that were 35.7% higher. Among people aged 65 and older in 2008, 27.0% of those in communities of color had incomes below the poverty line, compared to 10.8% of whites aged 65.

The Political Environment and Institutions

The City of Portland is governed by City Council, which consists of four city commissioners and a mayor. City commissioners are responsible for legislative policy and are assigned, by the mayor, a portfolio of Bureaus that are responsible for the city's operations. Each Bureau works closely with various partners, including other city entities (e.g., offices, commissions), governments (e.g., county and regional governments), the private sector, community groups, and educational institutions. Bureaus in Portland that have been actively involved in age-friendly efforts include Planning and Sustainability; Housing; Parks and Recreation; Transportation; Development Services; and Environmental Services. Additionally, the Office of Neighborhood Involvement, the Office of Equity and Human Rights, the Planning and Sustainability Commission, and the Portland Commission on Disability have played roles.

According to North Carolina State University (Healy, 2012), the Portland region and the Willamette Valley voted predominately for Democrats in the last election, while most of southern, central, and eastern areas of Oregon voted for Republicans. Portland and the state of Oregon are considered highly participatory places, with many efforts spearheaded by local community groups (Abbott, 2001); however, involvement from traditionally underrepresented groups, including older adults, has been an issue, as we found in our baseline study of Portland's age-friendliness (Neal & DeLaTorre, 2007a, b).

Services for an Older Population

Portland has been described as the capital of good planning (Abbott, 2001) and as the poster child for regional planning, growth management, and other innovative urban planning policies (Mayer & Provo, 2004). However, planning for an aging population was not a priority. The age-friendly efforts that have been underway since 2006, along with the efforts of nonprofit advocacy organizations such as Elders in Action and AARP Oregon, have assisted in drawing attention to this area.

The service environment for older adults in the Portland area is quite rich. However, few city services specifically address issues of aging, but many of the services are attuned to the needs of older Portlanders. The county (Multnomah) is responsible for aging services, administering Older Americans Act programs, and contracting with nonprofit organizations to deliver services, as well as participating in the development of the statewide Aging and Disability Resource Connection. Metro, the regional government, deals with long-range planning, green spaces, and recycling and garbage services. The state of Oregon has long been known for its innovations with respect to long-term care, providing a number of community-based care options aimed at helping Oregonians to remain as independent as possible.

Cultural Considerations

Portland is known for its approach to city-making; as urban historian Carl Abbott (2001, p. 1) opined, Portland "holds in tension the distinct values of environmentalism and urbanism." Abbott explained that Portland has numerous community development corporations, is strong in public involvement, and is well-known for its success in creating a compact city model that is studied and followed by others interested in place-making and creating good urban form. At the same time, the popular television program "Portlandia" has promulgated the idea that Portland is a place "where young people go to retire" and a place that embraces idiosyncratic culture. In fact, a common sight on Portland automobile bumpers and bicycle fenders is a sticker that reads "Keep Portland Weird," and some Portlanders take pride in the city being among the leaders in strip clubs and breweries per capita in the United States. Portland is also known as an active, outdoorsy place, with urban hiking and cycling among its most popular activities. This quirky reputation and the perception of being a city that attracts young people may, to some degree, work against efforts to focus on aging as an important policy agenda. However, according to Jurjevich and Schrock (2012), the urban and natural amenities of Portland appear to be pull factors for older empty-nesters and retired college-educated migrants.

Preceding Age-Friendly Work

The Institute on Aging (IOA) at Portland State University (PSU) (our academic home) has been the primary force behind Portland's age-friendly-related efforts. In 1989, IOA and Urban Planning faculty developed training materials for urban planners in Oregon to create livable environments for older adults, followed by the publication *Planning for an Aging Society* for the American Planning Association (Howe, Chapman, & Baggett, 1994). In 2006, the World Health Organization (WHO) approached the IOA to become a part of its Global Age-Friendly Cities Project. We believe that Portland was selected both because of its reputation for good planning and because of the IOA's research reputation and previous work in this area and in global aging issues. Although no funding for the study was available, the IOA viewed this as an excellent opportunity to build on our interests in global aging, urban studies, and planning and gerontology, to honor PSU's motto, "Let Knowledge Serve the City," and, most importantly, to improve Portland for future generations.

A total of 33 cities in 22 countries participated in the original WHO project (Portland was the only U.S. city; New York City joined later, during the data analysis phase). WHO's goal for the study was to identify concrete indicators of an age-friendly city and to produce a practical guide to stimulate and guide advocacy, community development, and policy change to make urban communities around the world age-friendly. In each city, researchers were to conduct eight focus groups: three with older adults, one with informal caregivers of older adults, and four with service providers and businesses (public, private, and nonprofit). The Portland project was guided by a team of advisors, including older adults and representatives from AARP Oregon, Elders in Action, and other public, private, nonprofit, and university organizations. The aim was to identify features of the city that were and were not age-friendly and to gather suggestions for improvement in the built, social, and service environments.

The findings from the Portland study were published in 2007 in a final report to the WHO (Neal & DeLaTorre, 2007a) and then in a summary of findings (Neal & DeLaTorre, 2007b), which was printed with support from AARP. These findings, along with those from each of the other participating cities, were used in crafting the WHO's *Global Age-Friendly Cities: A Guide* (WHO, 2007) and have been disseminated locally, nationally, and internationally.

STRUCTURE FOR THE INITIATIVE

The structure for the Age-Friendly Portland Initiative has evolved from the WHO age-friendly cities study in Portland, which provided the baseline assessment of Portland's age-friendliness and laid the groundwork for the IOA-City of Portland-Community partnership that exists today. The advisory group called together to guide the WHO study was expanded to the current Age-Friendly Portland Advisory Council, which is coordinated by the IOA. The following describes how this evolution occurred and who was involved.

Involvement of Civic and Political Leaders

In 2010, WHO launched its Global Network of Age-Friendly Cities (now "Cities and Communities") to provide a means to link cities for the purposes of information sharing, to ensure a common understanding of the "age-friendly" label, and to ensure that interventions undertaken toward becoming age-friendly would be appropriate, sustainable, and cost-effective. The IOA approached then Portland mayor Sam Adams with a request to submit an application for membership to the Network on behalf of the city. The mayor agreed, and the application was submitted. In addition, Mayor Adams asked us to join his Portland Plan Advisory Group (PPAG). The Portland Plan was a three-year action plan undertaken by the Bureau of Planning and Sustainability preparatory to revising and updating the city's 25-year strategic, Comprehensive Plan to make Portland a more thriving and sustainable city for all of its residents.

In June 2010, the IOA received notification that Portland had been selected as one of the initial nine-member cities in the WHO Global Network and advised the mayor. In a press release announcing Portland's membership, Mayor Adams noted that the city and its bureaus would work with the IOA to create the Network's required three-year citywide action plan in conjunction with the Portland Plan and proclaimed, "We are excited that the City of Portland has been chosen for membership in this first wave of WHO's Global Network of Age-Friendly Cities. As the mayor of Portland, I feel it is important that we embark on the process of creating a city that is friendly to people of all ages and abilities. . . . The timing of the Network's creation falls directly in line with the Portland Plan, and we feel that we are in a great place to move forward with a synergistic partnership. We look forward to a relationship with the WHO and other cities within the Network."

Despite IOA researchers attending numerous PPAG meetings and bringing attention to aging-related issues, the initial draft of the Portland Plan contained few explicit details pertaining to older adults. In response, we and partner agencies provided written suggestions to city staff, who subsequently requested a meeting to address the deficiencies identified in the draft plan. The final version of the Portland Plan contained a two-page section titled "Portland is a Place for All Generations," which called for the creation of an age-friendly action plan and other related actions. An action plan (now written and approved, as described in a subsequent section) was also required for continued membership in the WHO's Global Network. Its development was guided by the Age-Friendly Portland Advisory Council (AFP Council) with representation from key public, private, nonprofit, and university sectors. Liaisons to the AFP Council have been designated by each city commissioner and the mayor. Formal recognition of Portland's membership in the Network came in 2011, with the arrival of the official certificate of membership from the WHO, which was presented to Mayor Adams and City Council. At the time of this writing, the revision of the Comprehensive Plan is nearing completion, guided by the Portland Plan and clearly including consideration of the importance of building a city that works for people of all ages and abilities.

As detailed in a recent paper published in the *Journal of Aging and Social Policy* (Neal, DeLaTorre, & Carder, 2014), how age-friendly city concepts have come to be adopted as planning principles in Portland can be understood through Kingdon's (1984) theory of agenda setting. Kingdon's theory posits that a policy can be adopted or changed when a "window" of opportunity opens and permits participants in and outside of government to connect at least two of the three streams associated with the policy process. These streams include problem streams (i.e., defining and placing an issue on the agenda), policy streams (i.e., knowledge and solutions that can be considered by decision-makers), and political streams (i.e., the political climate and will to place the issue on the agenda).

Participation of Older People

The voices of older adults have been present in the project since the beginning through having representatives on the initial advisory group and now the AFP Council. Older adults and their caregivers were key participants in the baseline assessment, having attended presentations concerning the findings and provided feedback. Older adults have also been involved in town hall meetings to help identify priorities for action and in a mayoral candidate forum aimed at learning about candidates' perspectives and priorities with respect to planning for Portland's aging population. In addition, service-learning projects with older adults have been aimed at making particular features or areas of Portland more age-friendly (e.g., walkability audits). Finally, organizations represented on the Advisory Council, such as AARP Oregon and Elders in Action, provide opportunities for older adults to become involved in advocacy efforts and community projects.

Professional Staffing

The WHO project to assess Portland's baseline age-friendliness was conducted by the authors—the director of the IOA and a then-doctoral student and current research associate. The findings of the study were disseminated by us locally, nationally, and internationally through the IOA's website, holding meetings with city staff and community organizations (aging-related and not), sharing as members of civic committees, such as the PPAG and Portland Plan and Comprehensive Plan work groups, presenting at national and international conferences, and consulting with other cities. Several community partner organizations have participated actively by serving on the original advisory group and/or the current AFP Council and contributing to the development and implementation of the action plan. In December 2013, the IOA was joined by an Encore Fellow, recently retired from and stipended for one year by Intel Corporation.

Financial Resources

The financial resources available for Portland's age-friendly efforts have been minimal. Funding for the initial baseline study was provided through the IOA through in-kind hours from the director, meeting space, and tuition remission and a stipend for the doctoral student provided by the School of Community Health. Subsequently, some funding was provided through the Aging Matters, Locally and Globally Initiative established at the IOA by private donors, offering partial salary support and helping to cover the costs of travel to conferences to disseminate findings.

Funds for printing the summary of findings from the baseline assessment and a 2012 policy paper explaining the need for age-friendly planning and policymaking were provided by AARP Oregon. AARP Oregon also has covered town hall-related costs (e.g., room rentals, refreshments) and provided staff time for organizing the forums. Elders in Action has provided meeting space and staff time as well, as have the City's Bureau of Planning and Sustainability and all other community partners.

Other Important Structural Features

A central feature of the Age-Friendly Portland effort is that it has been, above all, a collaborative effort. The IOA carried out the original research for the WHO Global Age-Friendly Cities project, but the work could not have been accomplished without the aid of our community partner organizations and older adults themselves. Subsequently, the mayor's office, the Bureau of Planning and Sustainability, the offices of City Commissioners Fritz and Fish, and now those of Commissioners Saltzman and Novick, along with Metro and Multnomah County Aging and Disability Services, became involved. The role of the IOA at this point is that of the "backbone organization," as detailed by Kania and Kramer (2011) in their seminal article concerning collective impact. Kania and Kramer argued that large-scale social change requires broad cross-sector orchestration, not isolated actions by individual organizations, and that this collaboration requires sustained infrastructure. As they note, "The backbone organization requires a dedicated staff separate from the participating organizations that can plan, manage, and support the initiative through ongoing facilitation, technology and communications support, data collection and reporting, and handling the myriad logistical and administrative details, including the treasury, accounting and auditing functions that are needed for the initiative to function smoothly" (p. 40). The question facing the Portland age-friendly effort is how best to support this infrastructure moving forward.

Also central have been the two key values that have begun guiding planning in the Portland region—sustainable development and equity. These values permeate the Portland Plan and the revisions being made to the Comprehensive Plan and are inherently part of the Age-Friendly Portland initiative as well.

SUBSTANTIVE AGENDA: ACTIONS COMPLETED AND IN PROGRESS

Table 7.2 describes the actions completed since Portland's participation in the original WHO Global Age-Friendly Cities project and acceptance as a member in the WHO Global Network of Age-Friendly Cities and Communities.

Planning Underway

Once the Action Plan for an Age-Friendly Portland was approved by City Council, the AFP Council decided to prioritize activities in three focal areas for 2014–15: civic engagement, employment, and the economy, and housing. Subcommittees have been formed to facilitate the accomplishment of actions within each of these areas. Members include AFP Council members and others selected for their expertise and interest. Meetings with the city liaisons have been held and introductions made to city bureau and AFP Council lead staff. Interest in dementia-friendly environments has emerged, and collaborations are developing around this area. Research funding is being sought to help support the coordination of the initiative, implement particular actions, evaluate program outcomes, and inform future policy.

Although indicators to track progress with respect to the Action Plan have not been finalized, available data have been identified and analyzed. The actions in progress and those completed have been tracked by the AFP Council. Related age-friendly efforts and opportunities have also been catalogued to look for possible areas of collaboration and are being monitored. Work continues on developing indicators to monitor progress for each of the action steps outlined in the Action Plan. IOA researchers are engaged, as well, in national and international efforts to conceptualize and operationalize an approach to determine programmatic success and future directions (e.g., the development of indictors).

Table 7.2 Activities Completed in Conjunction with Portland's Age-Friendly Efforts

Policy Documents/ Plans	• Institute on Aging (IOA) researchers informed the Portland Plan, including a section aimed at making Portland a "place for all generations;" the Plan, approved by City Council in 2012, stipulated development of an age-friendly action plan • IOA researchers informed Neighborhood Centers policy language as part of the update of Portland's Comprehensive Plan; focused on enhancing aging in community and improving access to housing, transportation, and services (2013) • Action Plan for an Age-Friendly Portland unanimously approved by City Council resolution (October 16, 2013); Plan delineates 10 areas of action (the original eight World Health Organization (WHO) domains, but separating Civic Participation from Employment and Community Support from Health Services); the Mayor and City Commissioners have each designated liaisons to the Advisory Council
Reports	• Six Masters of Urban and Regional Planning students engaged Portlanders in envisioning an age-friendly Portland, created policy recommendations (2012) • A Portland State University (PSU) Institute for Portland Metropolitan Studies Fellow worked with IOA researchers to identify age-friendly indicators to track progress over time; he also created a data story (Greater Portland Pulse, 2012)
Publications	• Published article using geographic information system analysis to offer suggestions for increasing the region's age-friendliness (DeLaTorre et al., 2012) • Published article concerning aging as an equity issue for the Coalition for a Livable Future's Equity Atlas project (DeLaTorre & Neal, 2014). • Published paper describing how the age-friendly policy agenda was advanced (Neal et al., 2014)
Consultations	• Portland, coordinated by the IOA, participated in the WHO's pilot project to identify existing data sources for 61 possible indicators of age-friendliness (WHO, 2014); IOA researcher began serving on the WHO's Age-Friendly Indicators International Task Force • IOA researchers have collaborated with nearby counties on their initiatives to improve their environments for older adults, including Clark County (Washington), Clackamas County, and Washington County

(*Continued*)

Table 7.2 (Continued)

Projects	• Hands On Greater Portland (an affiliate of the United Way), Elders in Action, and AARP Oregon started an age-friendly service-learning program in 2012 to facilitate volunteer engagement in service projects (e.g., grocery delivery, walkability audits) • The Age-Friendly Portland Advisory Council launched www.agefriendlyportland.org to house information about the initiative and action plan, resources, and ways to become involved; the website was designed in conjunction with students from the Art Institute of Portland, is hosted by Elders in Action, and is administered by an Intel Encore Fellow at the IOA • Villages NW, the region's village movement, has worked with the Advisory Council to coordinate and disseminate information about both groups' efforts

Portland's age-friendly efforts have not directly led to the establishment of stand-alone political institutions. Instead, members of the AFP Council have worked with the mayor and the city commissioners to identify liaisons on their staff that attend the AFP Council meetings and facilitate Council work with staff from various city bureaus and offices. IOA researchers and other AFP Council members also worked with members of City Council to ensure that the Office of Equity and Human Rights, created by City Ordinance in 2011, would include addressing age as an equity issue in its scope of work.

OUTCOMES

Portland's age-friendly efforts to date have focused primarily on research, policy, and planning. However, several objective outcomes have been achieved. An AFP Council member worked with members of the Portland Commission on Disability and other community members to conduct an audit of city hall's accessibility. Based on the group's recommendations, improvements were made that increased accessibility above and beyond minimum standards. Additionally, in collaboration with community partners, members of the AFP Council have been successful in effecting improved transportation infrastructure in Portland, including design recommendations for portions of the South Waterfront where sidewalks, cycling paths, roads, streetcars, and the city's aerial tram intersect. Accessibility improvements have also been made to the regional rail system, with the suggestions to fix the location of the accessible light rail car and improve ramp design on streetcars being implemented.

Since Portland's participation in the WHO Age-Friendly Cities project in 2006, public awareness of both age-friendly features and barriers seems to have increased. Requests to join the AFP Council and to serve on action subcommittees have increased, and there has been a marked increase in the level of participation from public and nonprofit partners. Liaisons to elected officials, as well as city bureaus and offices, have helped to facilitate age-friendly efforts, including informing the Action Plan for an Age-Friendly Portland, the Portland Plan, and Portland's Comprehensive Plan. Additionally, AARP Oregon and Elders in Action have expanded their age-friendly programs, including age-friendly service-learning programs and coordinating volunteers to assess businesses' age-friendliness. Requests for audits of age-friendliness have increased from city bureaus and from businesses and nonprofit organizations. Furthermore, the general public has been involved in town hall meetings and forums aimed at improving Portland's age-friendliness.

Local, national, and international media coverage of these age-friendly efforts has come in the form of print, radio, and television coverage. We have received requests to serve on national and international task forces, invitations to speak, and invitations to consult on other age-friendly projects in the United States.

CONSTRAINTS

A number of economic, political, and cultural constraints still exist as limitations or even impediments to the Age-Friendly Portland initiative.

Economic

In a time of declining resources and increasing needs, economic constraints are a reality for both the funding of age-friendly programmatic efforts and of services and infrastructure that are age-friendly. Funding for Portland's age-friendly efforts has been almost entirely in-kind, with some funding from AARP Oregon for events, publications, and intern support. Grant funding has been difficult to secure based on the wide-ranging domains of age-friendliness and the constraints of many funding agencies. Funding from government agencies has been nearly impossible to obtain due to the economic conditions in the city. Additionally, frequent cuts to public transit and home-based care, combined with reductions in funding to Portland's neighborhood grants program, have reduced access to services and programs.

Political

Support for the age-friendly efforts in Portland has increased over the years, but the age-friendly policy agenda competes with other policy agendas, all of which have their merits. For instance, Portland's Cradle to Career initiative is aimed at improving educational opportunities for Portlanders, and the city's newly created Office of Equity and Human Rights has chosen to focus its efforts first on race and disability. Another political constraint has been turnover in elected officials and their staff and a resulting lack of continuity. There have been three mayors since the beginning of the WHO age-friendly project in Portland, some turnover in city commissioners, and major changes in commissioners' bureau assignments and day-to-day leadership with each new mayor. Furthermore, Portland's political landscape is complicated, as city, county, regional, and state politics all need to be considered and are, at times, disconnected and misaligned. For instance, major revisions to health care policy have been taken on at the state and county levels, while the planning, design, and development of infrastructure occur mainly at the city and regional levels. Embedding age-friendly concepts in local, regional, and state policy will help to overcome but will not eliminate these challenges.

Cultural

Portland has garnered a reputation as a place that attracts young, quirky, outdoor-oriented people. The perception of many is that Portland is a beacon for the young creative class. There is evidence that older people are moving to the region as well, but the potential for a cultural divide between young and old exists. Educating the community concerning the benefits of an aging population and of age-friendly environments for people of all ages and abilities is critically important.

IMPLICATIONS AND CONCLUSION

Efforts to make Portland a more age-friendly city have their roots in research and training efforts at the IOA in the late 1980s. Participation in the WHO's Global Age-Friendly Cities research project provided recent impetus and structure, particularly for the development of the Action Plan. The approach for creating a more age-friendly Portland has been policy-focused and long-term in nature, taking advantage of policy windows created with the strategic planning tools being implemented in the city, including the Portland Plan and the Comprehensive Plan. Age-friendly efforts have been infused into these policy documents, and the broader push in Portland to create a more sustainable and equitable place. This approach has resulted in a somewhat slow process of becoming more age-friendly but one that hopefully will pay benefits for years to come. Funds to coordinate efforts and to implement the various action steps detailed in the action plan would expedite the process.

REFERENCES

Abbott, C. (2001). *Greater Portland: Urban life and landscapes in the Pacific Northwest.* Philadelphia, PA: University of Pennsylvania Press.

Chatterjee, A., & DeVol, R. C. (2012). *Best cities for successful aging.* Milken Institute. Available from http://www.milkeninstitute.org/pdf/best-cities-successful-aging.pdf

City of Portland. (2009). *Portland plan background report: Housing supply.* Portland, OR: City of Portland Bureau of Planning and Sustainability. Available from http://www.portlandonline.com/portlandplan/index.cfm?a=270880&c=51427

City of Portland. (2012a). *The Portland plan.* Portland, OR: City of Portland Bureau of Planning and Sustainability. Available from http://www.portlandonline.com/portlandplan/

City of Portland. (2012b). *East Portland in motion: A five-year implementation study for active transportation.* Portland, OR: City of Portland Department of Transportation. Available from http://www.portlandoregon.gov/transportation/article/391907

City of Portland. (n.d.). Creating healthy, safe ways to get around Portland. Portland, OR: City of Portland Bureau of Transportation. Available from https://www.portlandoregon.gov/transportation/article/455963

Curry-Stevens, A., Cross-Hemmer, A., & Coalition of Communities of Color. (2010). *Communities of color in Multnomah County: An unsettling profile.* Portland, OR: Portland State University. Available from http://www.coalitioncommunitiescolor.org/

DeLaTorre, A., DeLaTorre, T., Neal, M. B., Carder, P., Weinstein, J., DeShane, M., & Wilson, K. B. (Winter, 2012). Periodic atlas of the metroscape: Planning for our aging society. In *Metroscape: An atlas and mirror of the Portland metropolitan region* (pp. 14–20). Portland, OR: Institute of Portland Metropolitan Studies, Portland State University. Available from http://www.pdx.edu/ims/metroscape-winter-2012

DeLaTorre, A., & Neal, M. B. (2014). Aging and equity in the greater Portland metropolitan region. *Connections, 12*(1), 10–23. Available from http://clfuture.org/equity-atlas

Greater Portland Pulse. (2011). *The path to economic prosperity: Equity and the education imperative.* Portland, OR: Portland State University and Metro. Available from http://www.portlandpulse.org/sites/default/files/reports/GPP_Report_0.pdf

Greater Portland Pulse. (2012). Age-friendly region. Portland, OR: Portland State University and Metro. Available from http://www.portlandpulse.org/age-friendly_region

Healy, C. (2012). *2012 U.S. election visualizations.* Raleigh, NC: North Carolina State University. Available from http://www.csc.ncsu.edu/faculty/healey/US_election/#OR

Hammond, B. (2009, January 17). In a changing world, Portland remains overwhelmingly white. *The Oregonian.* Available from http://www.oregonlive.com/

Howe, D. A., Chapman, N. J., & Baggett, S. A. (1994). *Planning for an aging society* (Planning Advisory Service Report Number 451). Chicago, IL: American Planning Association.

Jurjevich, J., & Schrock, G. (2012). *Is Portland really the place where young people go to retire? Migration patterns of Portland's young and college-educated, 1980–2010*. Portland, OR: Portland State University. Available from http://mkn.research.pdx.edu/wp-content/uploads/2012/09/JurjevichSchrockMigrationReport1.pdf

Kania, J., & Kramer, M. (2011, Winter). Collective impact. *Stanford Social Innovation Review, 56*, 36–41. Available from http://www.ssireview.org/

Kingdon, J.W. (1984). *Agendas, alternatives, and public policies* (2nd ed.). New York: Harper Collins College.

Mayer, H., & Provo, J. (2004). The Portland edge in context. In C. P. Ozawa (Ed.), *The Portland edge: Challenges and successes in growing communities* (pp. 9–34). Washington, DC: Island Press.

Metro. (2009). *20 and 50 year regional population and employment range forecasts*. Available from http://library.oregonmetro.gov/files/2030–2060_forecast_april_09.pdf

Metro. (2014). *Active transportation program*. Available from http://www.oregonmetro.gov/index.cfm/go/by.web/id/30078

Neal, M.B., Chapman, N., Dill, J., Sharkova, I., DeLaTorre, A., Sullivan, K., . . . Martin, S. (2006). *Age-related shifts in housing and transportation demand: A multidisciplinary study conducted for Metro by Portland State University's College of Urban and Public Affairs*. Portland State University. Available from www.pdx.edu/sites/www.pdx.edu.ioa/files/media_assets/cupa_age_related_shifts.pdf

Neal, M.B., & DeLaTorre, A. (2007a). *The World Health Organization's age-friendly cities project in Portland, Oregon: Final report*. Portland State University. Available from http://www.pdx.edu/sites/www.pdx.edu.ioa/files/media_assets/ioa_age_friendly_cities.pdf

Neal, M.B., & DeLaTorre, A. (2007b). *The World Health Organization's age-friendly cities project in Portland, Oregon: Summary of findings*. Portland State University and AARP. Available from Portland State University website: http://www.pdx.edu/sites/www.pdx.edu.ioa/files/ioa_age_friendly_cities.pdf

Neal, M.B., DeLaTorre, A.K., & Carder, P.C. (2014). Age-friendly Portland: A university-city-community partnership. *Journal of Aging and Social Policy, 26*(1–2), 88–101.

Portland Business Alliance. (2010). *A check-up on the Portland-region's economic health*. Portland: OR: Portland Business Alliance. Available from http://www.orbusinesscouncil.org/documents/portlandregioneconcheckup.pdf

TriMet. (2014). *TriMet in the news*. Available from http://trimet.org/about/inthenews.htm

U.S. Census Bureau. (2005). *Interim state population projects: 2005*. Available from http://www.census.gov/population/projections/data/state/projectionsagesex.html

U.S. Census Bureau. (2008). *National population projections: Summary tables*. Available from http://www.census.gov/population/projections/data/national/2008/summarytables.html

U.S. Census Bureau. (2012). *The 2012 statistical abstract: The national data book*. Available from http://www.census.gov/compendia/statab/cats/population.html

U.S. Census Bureau. (2013). *Estimates of population change for metropolitan statistical areas and rankings* [Data file]. Available from http://www.census.gov/popest/data/metro/totals/2012/

U.S. Census Bureau. (2014). Portland (city), Oregon [Data file]. *State & County Quick Facts*. Available February 1, 2014, from http://quickfacts.census.gov/qfd/states/41/4159000.html

World Health Organization (WHO). (2007). *Global age-friendly cities: A guide*. Geneva, Switzerland. Available from http://www.who.int/ageing/publications/

World Health Organization (WHO). (2014). *2nd WHO consultation on developing age-friendly city indictors, 8 September 2013, Québec City, Canada*. World Health Organization, Center for Health Development. Available from http://www.who.int/kobe_centre/ageing/age_friendly_cities/AFC_Quebec-mtg_SEP13/en/index.html

8 The Environment of Environmental Change

The Impact of City Realities on the Success of Age-Friendly Programs

Allen Glicksman and Lauren Ring

INTRODUCTION

The goal of most "age-friendly" efforts is to transform parts of the social and physical environments to improve quality of life for older adults (WHO, 2006). We believe that the success or failure of these attempts to modify the environment depends on the larger social, economic, political, and physical contexts in which these changes occur. This case study uses age-friendly efforts in Philadelphia to illustrate our point. We focus on two key elements in the larger city environment—the decline in working class jobs and the resulting decline in the size of the city's population, on the one hand, and growing efforts to renew the city at the neighborhood level, on the other—as critical to the outcome of local age-friendly efforts. Although there are some unique aspects of the Philadelphia story (such as the presence of two age-friendly efforts), many of the challenges facing the city are the same as those facing other large American cities, especially in the northeast part of the country.

There are two key lessons that other age-friendly efforts can learn from this case study. First, that age-friendly efforts must remain mindful of the context in which they occur. As we will show, the age-friendly efforts in Philadelphia are dependent on and in turn influence other changes going on in the city. Second, to be truly effective, age-friendly efforts must build strong connections with persons and organizations outside the network of aging services who are working toward the same general goals of community strengthening and environmental change. Only by building alliances with others committed to similar changes at the neighborhood and city levels can age-friendly efforts reach their goals.

HISTORICAL BACKGROUND

Philadelphia was established in 1682 by William Penn. By 1750, it was the largest port in the American colonies. Its role as a major port encouraged the growth of various industries in the city. The processes of industrialization led to a series of conflicts among various groups within the city, conflicts that were exacerbated by growing ethnic, religious, and racial tensions. At times these tensions flared into riots, most notably the nativist riots of 1844 that pitted Irish Catholics against Protestants. These riots are an early example of the continuing underlying ethnic and racial tensions in Philadelphia that continue to play a significant role in shaping the life of the city to the present day (Warner, 1969).

Attempts at "urban renewal" (a term not used until the 1950s) were already under way by the end of the nineteenth century. Philadelphia, along with other major urban areas, focused much of this effort on the development of a large railroad station to become a railroad hub. In Philadelphia, it was hoped that building this new station would help to ". . . clear out deteriorated buildings and detested residents" by tearing down low-income housing to make room for the train station (Avila & Rose, 2009, p. 336). Those "detested residents" were often poor members of racial and ethnic groups despised by many of the city's leading citizens. The goal of improving the city by making it more attractive to business, on the one hand, and the removal or isolation of unwanted housing and residents, on the other, continued to characterize urban renewal efforts in Philadelphia into the twentieth century. Efforts at urban renewal intensified after the Second World War. These post-war efforts were often characterized by linking housing redevelopment, slum clearance, and urban expressways (the expressways having often replaced trains as the method of choice for the wealthy to travel to and from work) (Avila & Rose, p. 338). These efforts created greater geographic dispersion among the population across the region and less mixed land use. This separation of residential areas from commercial and manufacturing areas was a hallmark of urban renewal throughout the decade.

The period after the Second World War also marked the beginning of a decline in population due to changes in the economic base of the city. Philadelphia lost 18.5% of its population between 1950 and 1980. Between 1953 and 1982, 36.4% of its non-agricultural business establishments and 24.9% of its jobs disappeared (Beauregard, 1990, p. 423). Much of this decline reflected changes in technology that eliminated the need for many unskilled positions. This was especially true of the way cargo was moved at the Port of Philadelphia, a change that occurred in many port cities around the country. The jobs lost were often working-class jobs as manufacturing businesses began either to close or to leave the city (Beauregard, 1990).

Endeavors at reversing the trend of the loss of working-class jobs became intertwined with efforts to prevent entire neighborhoods from falling into poverty. These types of efforts, tying together citywide economic recovery with individual neighborhood revitalization, became more common in the 1970s. At that time, federal funding for urban renewal was shifting to provide greater decision-making to local politicians and a commitment to a "color blind" process in an attempt to remove race as a factor in deciding how monies were spent (Avila & Rose, 2009). The concept of "urban renewal" itself underwent a change, from being a top-down effort to remove certain groups and neighborhoods from the city core to an attempt to revitalize precisely those neighborhoods with the highest proportion of poor and minority residents. The idea of a "color blind" process was rarely realized in Philadelphia, where race continued to be a contentious issue in local politics (Ley, 1974). This was exacerbated by a continuing outflux of white residents from the city to suburbs in Pennsylvania and New Jersey, often spurred by the economic and political problems the city faced. In fact, in Philadelphia and other large cities, race and class issues seemed impossible to separate. At the same time, local politics became dominated by conflicts about growth, as the city experienced a weakening tax base and increased underemployment. These efforts continued through the 1980s. In particular, there was an attempt to redress problems in equality by reforming government and using race as a base for hiring for city contracts (Beauregard, 1990).

The long-term impact of the decline in working-class jobs can be seen in the list of the 10 top employers in the city today. Of the 10 largest employers in the city, three are government (Federal government, City of Philadelphia, and the School District of Philadelphia); five are medical/academic centers (University of Pennsylvania, Children's Hospital of Philadelphia, Thomas Jefferson University Hospital, Temple University, and Albert Einstein Medical Center); and two are transportation (South Eastern Pennsylvania Transportation Authority and U.S. Airways). There are few manufacturing, industrial, or similar positions available in these types of institutions, and many of the positions require at least a B.A. degree or equivalent, as well as work experience in related positions (Technical.ly/Philly, 2013).

By the early 1990s, the fiscal problems that faced the city had caused the state legislature to establish a fiscal oversight board. The mayor elected in 1991, Ed Rendell, balanced the budget and also worked on improving services to city neighborhoods. Rendell's successor, Mayor John Street, started a $296 million Neighborhood Transformation Initiative. The initiative was a major element of Street's election campaign, although it ran into difficulties due to conflict with City Council. This initiative included a plan to make the city more environmentally friendly and to involve residents in some of the decision-making (Riordan, 2006).

At this time, some neighborhoods began to gentrify. In some cases, neighborhoods went through the more common form of gentrification, turning from having primarily low-income residents to having high-income residents. In other neighborhoods to which significant numbers of well-educated, upwardly mobile young people moved, average incomes did not increase by much, but the difference in socioeconomic status between the long-term residents and the new arrivals was readily apparent and sometimes led to tensions between the two groups. The continuing gentrification of parts of the city has not stopped an increase in the number of Philadelphians living in poverty, especially deep poverty (less than 50% of the poverty level) (Lubrano, 2013).

Residents of some low-income neighborhoods began to seek ways of improving the areas in which they lived and to gain a voice in the process of deciding how and when neighborhood change would occur. In some cases, residents were concerned about gentrification, especially when it seemed to be driven by expanding universities in the city (Brownlow, 2011). At the same time that universities could be seen as a danger to long-existing neighborhoods, some of those efforts involved university activists in building their case for neighborhood change. In one case, a model was developed to use research from a university to support a neighborhood group using data to demonstrate the relation of blight to population loss. That same research was used by other groups to work for change. The growth of neighborhood-based organizations, such as Community Development Corporations, which worked to revitalize challenged neighborhoods, also signaled a new approach to urban renewal (Young, 2001). More recently in Philadelphia, perhaps due to a renewed nationwide interest in the benefits of urban living, there has been an influx of young professionals in their 20s and 30s. Many of these individuals are working or volunteering at city agencies, local nonprofits, civic associations, and school districts and would serve as an unexpected resource for age-friendly efforts in the city (Wieckowski, 2010).

AGING IN PHILADELPHIA

The changes in the city population that occurred over the decades preceding the beginning of age-friendly efforts can also be seen in changes to the size and composition of its older adult population. As eligibility for most aging programs in the city begins at age 60, we are using that cut-off for the definition of who is counted as an older Philadelphian. In 1980, there were approximately 326,000 older adults in the city. By 2009, when the first Age-friendly Philadelphia (AfP) effort began, that number had dropped to 263,000. However, the older adult population in the city had not just declined in number during these years, it had also been transformed. Over the same period, from 1980 to 2009, the proportion of white, non-Hispanic American-born persons among the city's elders dropped from 58% to 45%. While the proportion of foreign-born held steady at about 15%, the proportion of those foreign-born from Europe was cut in half form 80% to 41%. The proportion from Latin America doubled from 7% to 15%, and the proportion from Asia and the Pacific islands went from 2% to 22%. So, over this 29-year period, as the city's elders declined in number, the city increased in diversity (Glicksman, 2014, unpublished).

At the same time that the population of older adults declined and the diversity increased, some characteristics of the older adult population remained steady—most importantly among those characteristics, poverty. Over the same 29-year period, the proportion of older adults in the city with incomes less than 100% of the federal poverty level rose slightly from 18% to 22%. Most strikingly, the proportion of single-person households went from 28% to 38% in the same time period, so isolation and poverty remain key risk factors for older adults living in the city. By the time the first AfP effort was launched, Philadelphia had the second largest proportion of older adults of the ten largest cities in the United States after New York, one of the first American cities to develop an age-friendly effort. More than half were minority or foreign born and almost half were low income.

AGE-FRIENDLY PHILADELPHIA

It was within this context of growing poverty and attempts to transform the city at the neighborhood level that the city's two age-friendly efforts were established. The first of these AfP efforts was initiated in 2009 by Philadelphia Corporation for Aging (PCA), growing out of existing programs in PCA's Research and Planning departments. PCA, established in 1973, is the Area Agency on Aging (AAA) for the City of Philadelphia. AAAs were established by the federal government to distribute funds allocated through the Older Americans Act passed in 1965. The programs and services offered under that act are designed to help older adults remain in their homes and communities for as long as possible.

The second AfP effort, led by the City of Philadelphia, was established two years later and was born out of a strategic plan designed by the Mayor's Commission on Aging (MCOA) in order better to serve the changing needs of the city's elders.

Background: PCA and AfP

From its inception, PCA has dealt with an increasingly poor and increasingly ethnically and racially diverse older adult population. By the early part of the twenty-first century, the increase in the diversity of the population, especially the growing numbers of immigrants and refugees among the city's elders, encouraged the agency to add two new programs to the agency—Community Relations, which offered outreach to various ethnic and limited English-speaking populations, and Research, whose purpose was in part to help the agency better understand these new populations. Much of the work done by the PCA AfP effort has been described in three previous articles (Clark, 2011; Clark & Glicksman, 2012; Glicksman, Clark, Kleban, Ring, & Hoffman, 2014). The effort was initially undertaken without reference to the recent history of the city described in the first part of this chapter. However, that recent history had a profound impact on the progress of the AfP effort.

An important goal of PCA's Research Program was to identify topics of importance regarding the city's older adult population, conduct research on these topics, and then share scientific findings with the planning, policy, and service organizations serving older Philadelphians. From the beginning, a focus on the impact of the urban environment was part of the research agenda, and one of the first major efforts of the Research Program was to share information about urban aging with other professionals in the city. In 2003, the Research Program sponsored the first M. Powell Lawton Conference on Urban Aging. The Lawton conferences, organized in memory of M. Powell Lawton, one of the founders of the field of environmental gerontology, bring together policymakers, planners, social workers, and others who provide direct service to older adults, with researchers to investigate topics related to urban aging. These conferences give public view to issues of concern across these professional groups. Many of the speakers do not have backgrounds in aging; speaking at the conference becomes a vehicle for their interests to be shared with the aging network and for them to learn more about the concerns of the aging network. Research Program staff at PCA also began presenting at academic conferences, in policy settings, and to social service organizations on the impact of the urban environment on the health of older Philadelphians. In addition, a new Planner was hired who had worked in housing and had extensive experience working with Community Development Corporations (CDCs). For the first time, the research agenda in urban aging was working with a planning agenda focused on community change.

The PCA AfP Effort

In 2008, the Planner who had initiated the work with CDCs left and a new Planner was hired. It was the new Planner, who came with a background in public policy, who first thought of starting an "age-friendly" effort in Philadelphia. Based on her previous experience working with nonprofits and city government, she developed an effort linking these types of institutions to an effort to improve environments for older adults. It was that experience in city government outside of Philadelphia and not the realities of life in the city that led her to explore those links. Seeking a framework for this effort, she began reading about "age-friendly" efforts in other cities. Based on her investigation of the emerging "age-friendly" movement, she selected the framework offered by the Environmental Protection Agency to adopt in Philadelphia (U.S. EPA, 2009). The EPA framework places emphasis on four principles; 1) Staying Active, Connected, and Engaged; 2) Development and Housing; 3) Transportation and Mobility; and 4) Staying Healthy (U.S. EPA, 2009). The EPA model had the advantage of defining four areas of effort that were of interest to both PCA and those involved in the most recent urban renewal efforts in the city, although those working on the AfP effort were unaware of that. Just as important, the EPA model brought together the concepts of Active Aging with Smart Growth, allowing for the integration of concepts from the field of aging with concepts from urban planning.

The selection of the EPA model led to a process of seeking out organizations and agencies outside the traditional aging network whose work affected the lives of older adults. The goal of meeting with these organizations (there were more than 150) was to link the aging services network with other networks, including environmental advocates, neighborhood organizations, food access and transportation initiatives (the latter two were already aging concerns), and even animal welfare groups. Many of these groups had not thought about the number of older adults in the city or how many older adults were to be found among their clients, volunteers, and donors. Some of these groups had been established as part of the effort by individuals and organizations in the city to encourage environmentally friendly growth. Beyond those organizations, many people in the city had not thought about the social roles played by older Philadelphians, including serving as the "the eyes and ears" of many neighborhoods and, in some cases, raising grandchildren and great-grandchildren. The Planner found ways of linking the four elements of the EPA framework into new or ongoing city efforts that were otherwise not tied to specific aging initiatives or services. For example, the AfP agenda item of increasing social capital was expressed by encouraging the implementation of an "age-friendly parks" program in the city. We were able to link that effort with park-related programs sponsored by a not-for-profit agency. As AfP and the parks program were committed to involving residents living near parks in developing more accessible parks, the two initiatives were able to support one another (Fairmount Parks Conservancy, 2014). The AfP program also took advantage of other changes going on in the city, so while AfP was aware of these changes that might be tied to the goal of making the city age-friendly, it was unaware of some of the enduring conflicts that made certain changes difficult. An excellent example of this is AfP's partially successful attempt to add age-friendly elements to the rewriting of the city zoning code. The code, which had not been revised in 50 years, was totally rewritten to reflect contemporary realities and helps to improve the development of new business and industry and to revitalize neighborhoods. The rewriting of the code occurred in collaboration with an initiative to create new City Planning Districts and develop a District Plan for each one to improve the neighborhoods. AfP worked with the commission rewriting the city zoning code and also encouraged other organizations that serve older adults in the city to express support for these additions to the code. What surprised the Planner and Research Director who were working with the Zoning Commission was the strong negative reaction to some of the proposed changes. When the concept of adding Accessory Dwelling Units (ADUs) to the code was first suggested, it seemed a benign idea. ADUs are created by subdividing a residence so that there are separate entrances to each unit. Sometimes called "granny flats" or "mother-in-law suites," it often allows an older adult to share the same building as a child, or allows an older adult to offer a rental space in his/her home. At neighborhood meetings across the city, the complaint was heard from residents that the universities would use this to gain control over more residential properties and push them out. The "town vs. gown" conflict that had been part of the struggle over urban renewal in the city for decades had suddenly intruded itself on the AfP effort. In the end, a very modest ADU section was added to the new code.

At the same time, AfP continued to engage in successful efforts with environmentalists and transportation advocates on projects such as age-friendly bus stops and encouraging community gardens at senior centers and senior housing. The effort at gardening brought AfP into contact with a growing number of groups in the city that had already begun working on reducing the number of "food deserts" and increasing access to healthy foods for persons of all ages.

The PCA AfP effort both responded to and took advantage of the challenges facing the city and attempts to improve it, as described above. First, the high rate of poverty and the large number of minority elders meant that interventions would have to focus on ways of helping elders with very basic needs, such as food and transportation. Any intervention that relied on the presence of significant financial resources or high rates of literacy among older adults would not benefit many in the city's older adult population. Second, given the history of some of these neighborhoods, efforts had to be based in local community organizations, such as community development corporations, and not in universities or other institutions viewed with suspicion by at least some of the city's older adult population.

On the other hand, the presence of efforts to improve the city overall, such as targeted efforts to retain more recent college graduates in Philadelphia or create a new Zoning Code, provided opportunities to link to AfP efforts. By tying AfP goals into city goals, such as ensuring that more older adults remain in their homes because they often served as the "eyes and ears" of the neighborhood, the goals of the AfP and the desire of the city leadership could be blended into a single set of efforts. Both of the AfP efforts have linked to neighborhood revitalization efforts, including efforts to improve parks and public places, for the benefit of all the residents, young and old.

Building Support Among Younger Professionals

Two special foci characterized the AfP effort. First, a great many of the organizations that met with age-friendly staff were represented by persons in their 20s and 30s (as was the Planner). Following up on some of these new contacts, a new organization, GenPhilly, was launched. GenPhilly was established in 2009, along with the AfP program. GenPhilly is described as ". . . a network of emerging professionals who are inspiring Philadelphians to connect with older adults through their work and personal lives. Through creative programming and professional development opportunities, we are working to change stereotypes and establish Philadelphia as a lifelong community for all of us" (GenPhilly, 2009). GenPhilly is unlike other groups formed to support aging issues because its membership reaches out beyond the aging network and incorporates members from government, the nonprofit sector, environmental groups, and other types of organizations. Their activities have included events on social capital and on neighborhood impact on older adults. They have also developed programs around senior gardening, transportation access, and technology issues. GenPhilly has been a critical element in the success of the AfP effort, bringing in people who would otherwise have little or nothing to do with aging and by sharing expertise in critical areas (such as transportation) that allow for the development of new, innovative ideas regarding access for senior adults. While at first we were surprised to see so many younger persons involved in the process of making the city more livable, we also became aware that one reason these positions were often held by younger persons is that many people, once they began to have children, moved out of the city, seeking better public schools and safer streets. This has occurred among some of the first generation of GenPhilly members as well. A fuller description of GenPhilly recently appeared in print (Clark, 2014).

The Role of Research in AfP

The second special focus was that, in spite of this being a policy effort, research was intimately involved in this process, as it had been from the beginning. This benefited the building of the age-friendly effort in two ways. First, a robust effort to develop in-house research that supported the goals of AfP was launched. Second, offering data, maps, and other support to collaborators in the effort helped to strengthen ties between AfP and other city-based agencies. Many of these agencies were applying for grant funding to extend and expand their own efforts. Several agencies mentioned this and added that some of the funding agencies were now looking for measurable outcomes and that often the funders specified health outcomes as the goal they wanted measured. As AfP could be helpful in developing appropriate, measurable health outcomes, this became another base for collaboration.

PCA's Research Program's effort on behalf of AfP began with an operationalization and testing of the EPA age-friendly model on which the effort is based. To test the potential impact of this model, the Research program decided to examine the relation of the four elements of the framework to health outcomes, while controlling for the impact of poverty and minority status. The creation of a formal model was important because the original wording of the EPA framework was very broad and often not specific regarding the interventions that might be appropriate for a given part of the framework. For example, the EPA item "Staying Active, Connected and Engaged" became Social Capital in AfP. By creating a formal model, we were able better to focus the general concept on a particular effort (being active and connected in one's neighborhood), and we could operationalize that concept through measures of Social Capital (Norstrand, Glicksman, Lubben, & Kleban, 2012). The Research Program also conducted focus groups to examine attitudes toward gardening and toward using city parks by older Philadelphians (Wang & Glicksman, 2013). The Research Program received an exploratory grant (Walkability's Impact on Senior Health, WISH) from the National Institute of Nursing Research to test the relation of aspects of the environment to health behaviors among older persons in Philadelphia (Glicksman, Ring, Kleban, & Hoffman, 2013). Further research has examined the impact of environment on need for services for older adults and linking the age-friendly environment effort to the use of the formal service network (Ferris, Glicksman, & Kleban, 2014). This part of the research agenda examined the impact of environment when minority status and poverty are accounted for in the analyses and determined that environment is still a significant predictor of health outcomes even when these factors are taken into account. What we did not expect to find, but was revealed in our analyses, was that age does not play as big a role as we predicted in the relation of health and environment—environment seems to have a similar effect on adults of all ages (Glicksman et al., 2013). Demonstrating the impact of environment on all Philadelphians allowed us to help agencies seeking environmental change not focused on aging.

In addition to internal research, the Research Program made a special effort to provide data to many outside agencies that became involved in the effort, as well as information to support elements of the effort, such as changes to the zoning code, and to develop links at local universities. Much of the research effort was focused on finding ways to present the results of analyses in a manner that made those findings useful for planning, policy, and advocacy. Two interrelated ways were found to accomplish that goal.

First, the use of Geographic Information Systems (GIS) played a critical role in this project both as part of the scientific research and as a tool for translating the scientific results into policy. Using geographically based data created in GIS in our analyses, we were able to demonstrate the critical role of environment on health outcomes. When we presented our findings to policymakers and planners, as we did at the Design Charrette (see below), maps of the neighborhoods being discussed were much more meaningful to the staff at these agencies than any sets of statistical tables we could have created. Maps, supplemented by tables, became the main way that information from the Research Program was shared with agencies (Ring, Glicksman, & Kleban, 2012). Although we were unaware of the effort, this mirrored the effort by researchers earlier in the decade to provide community activists with information to support their efforts at improving neighborhoods (Shlay & Whitman, 2006).

Second, a further challenge was to find ways to share research findings with community members and older adults along with the professionals who work with these populations. We discovered one effective method to accomplish this goal—holding a Design Charrette, an event or series of events that brings together community stakeholders to work collaboratively in community planning and design. What we found especially appealing was the way in which the method allowed professionals, technical experts, and community members to work together as equals in this process. As part of the WISH grant, a Charette was held with PCA and Asociación de Puertorriqueños en Marcha (APM), a Health, Human Services, & Community Development Organization based in one of the city's most challenged neighborhoods. The research team from the WISH grant participated, along with planners, architects, representatives of government agencies, social service professionals, and older residents of the lower north Philadelphia community (ThirdAge, 2012). This Charette was designed to introduce all parties to the findings of the WISH study and to gain input on what features of the neighborhood could be improved to promote health and well-being. The findings were presented using maps of the area that identified many of the characteristics that the study had identified as correlates of healthy behaviors. That led into a further discussion, bringing together the findings from the study and what the groups defined as the characteristics of a healthy neighborhood. That discussion resulted in specific recommendations for APM regarding ways to improve the neighborhood. Integrating research and planning through this process was invaluable to smooth the transition and facilitate the communication between the practice, policy, and planning networks.

Recognition of PCA's AfP Effort

In 2011, PCA's AfP effort received the EPA's Commitment Award for Building Healthy Communities for Active Aging; it also received a National Associations of Area Agencies on Aging "Aging Achievement Award" and was covered in several national and local media outlets. These recognitions of the impact of the AfP effort confirmed the importance of the work being done by all the agencies and individuals collaborating on this effort.

The City of Philadelphia's AfP Effort

The second age-friendly effort was established two years after PCA began its effort. While various city government offices had supported the PCA AfP effort, a decision was made that the city itself should show leadership in this area. In 2011, the Mayor's Commission on Aging (MCOA) created a Strategic Plan to make the city more age-friendly (Mayor's Commission on Aging, 2014). The focus of the plan is to help ensure that seniors can access resources critical for them to remain in their homes and neighborhoods, as well as to provide opportunities for them to engage with and contribute to their communities. This plan was the product of a reinvigorated MCOA that the Mayor created to better serve the changing needs of the city's elders. It also paralleled other city efforts, in areas such as volunteerism and environmental change, to influence both city government and other agencies to focus on these concerns. The Plan established five key goals for the MCOA: 1) serve as a catalyst and convener for new efforts on behalf of elders; 2) advocate for leveraging city resources; 3) educate all citizens about older adults; 4) promote opportunities for older adults to contribute to the city; and 5) encourage support for older Philadelphians to remain in their homes. This has lead to an application to the World Health Organization Global Age-Friendly Cities Program to join the Global Network of Age-Friendly Cities. The Mayor's Officer of Policy Planning and Development published the assessment in 2012 and is currently working on a follow-up study to better understand what older adults want from an age-friendly program. In addition, the MCOA has completed two reports regarding food access for older adults in the city. While there has been no formal division of labor between the two AfP efforts, some informal sharing of tasks has emerged. For example, PCA has taken the lead in AfP's research and evaluation efforts and has been active in supporting the MCOA effort with data, maps, and other resources. The MCOA has taken a leading role in bringing city government departments into the effort and into shaping a city-wide plan based on the WHO model. It is impossible to predict what the future relation between these efforts may be, or whether the two may be merged. Changes in the leadership in City government, as well as the priorities of PCA, may lead to rethinking not only the roles of the two efforts but how each effort supports the wider goals of both PCA and the City of Philadelphia.

AGE-FRIENDLY IN THE WIDER PHILADELPHIA CONTEXT

When AfP was launched in 2009, none of the staff involved was aware of much of the history of urban renewal in the city. While it was clear that the city had a very large number of older adults and that almost half were poor and more than half were minority, foreign-born, or both, the specific history of attempts to deal with the consequences of declining city fortunes was not considered as the framework when the effort was developed. As described above, the framework was instead taken from a model developed by the EPA for general use. But, unknowingly, AfP rediscovered what the advocates and partisans of neighborhood renewal in the city already knew: poverty and distressed neighborhoods were taking their toll on the residents of the city no matter what their ages. Many of the organizations and individuals who became active in GenPhilly were already committed to programs and initiatives designed to directly have impacts on the city's environment, initiatives that were the outgrowth of the two decades of efforts to reverse the declining tax base in the city and to attract a new generation of entrepreneurs, community leaders, and others who would revitalize the city's economy and infrastructure. The PCA AfP effort has had success in three main areas, including effecting modest changes to policies and programs that affect older adults, introducing aging as an issue to those outside of the traditional aging network, and making sure that aging is part of the ongoing agenda for the city and other agencies. One example that highlights these three areas is the effort to add aging concerns to the city's new zoning code. First, actual changes were made to the code. Although not as comprehensive as desired by aging advocates, some changes were introduced to benefit older adults. Second, key players in the zoning code effort were introduced to the topic of aging and the importance of thinking about older adults in the city. Finally, aging is now an agenda item for future discussions about the city. It is, in fact, probable that AfP only succeeded because these persons saw the same basic challenges and sought the same types of solutions. So the entire issue of ADUs and Visitable Homes would not have arisen if it weren't for the recognition that the city needed a new zoning code to enhance economic development. Thus, AfP added an aging component to an ongoing effort and, in doing so, benefited both elders and the city as a whole. It is not a coincidence that the citywide efforts and the age-friendly initiative both predate the first Census in decades (2010) that showed no decline in the population of the city. The city's embrace of AfP may also be ascribed in part to a more general commitment to making the city more livable. The city's AfP effort has made the entire issue much more visible. One example is the interest that Philadelphia City Council has now taken in the issue of aging, exemplified by the "Creating a City for All Ages" conference sponsored by a member of City Council in May 2011. It is noteworthy that, while both the MCOA and PCA were sponsors of that conference, the conference itself was not developed by either AfP effort. Rather, the presence of these efforts has put aging on the agendas of many other organizations.

"Age-friendly" resonated with many people because it offered a way to deal with problems faced by the city's elders in a manner that fit into similar efforts to transform the urban landscape. By creating efforts parallel to other efforts in the city, and collaborating with them, AfP brought aging to the agenda of organizations with no previous interest in aging. AfP enhanced and added to these citywide efforts. But none of this would have been possible if the age-friendly effort had not focused on the same issues that were at the top of the agendas of others, and offered solutions that were complementary to the other efforts occurring in the same period.

REFERENCES

Avila, E., & Rose, M.H. (2009). Race culture, politics and urban renewal: An introduction. *Journal of Urban History, 35*(3), 335–347.

Beauregard, R.A. (1990). Tenacious inequalities: Politics and race in Philadelphia. *Urban Affairs Review, 25*, 420. doi:10.1177/004208169002500305

Brownlow, A. (2011). Between rights and responsibilities: Insurgent performance in an invisible landscape. *Environment and Planning A, 43*, 1268–1286.

Clark, K. (2011) *Laying the foundation for an age-friendly Philadelphia: A progress report*. Philadelphia, PA: Philadelphia Corporation for Aging.

Clark, K. (2014). GenPhilly: A strategy for improving the sustainability of aging in community initiatives. *Journal of Aging & Social Policy*, *26*(1–2), 197–211. doi: 10.1080/08959420.2014.854135

Clark, K., & Glicksman, A. (2012) Age-friendly Philadelphia (AfP): Bringing diverse networks together around aging issues. *Journal of Housing for the Elderly*, *26*(1–3), 290–307.

Fairmount Park Conservancy. (2014). *Hunting Park Revitalization Project*. Available from http://myphillypark.org/what-we-do/capital-projects/hunting-park

Ferris, R., Glicksman, A., & Kleban, M. (2014). Environmental predictors of unmet Home- and Community-Based Service (HCBS) needs of older adults. *Journal of Applied Gerontology*, 1–30. doi:10.1177/0733464814525504

GenPhilly. (2009). *GenPhilly Newsletter*. Available from http://www.genphilly.org

Glicksman, A. (2014). Analysis of aging trends in Philadelphia using the American Community Survey. Unpublished raw data.

Glicksman, A., Clark, K., Kleban, M., Ring, L., & Hoffman, C. (2014). Building an integrated research/policy-planning age-friendly agenda. *Journal of Aging and Social Policy*, *26*(1–2), 131–146. doi:http://dx.doi.org/10.1080/08959420.2014.854142

Glicksman, A., Ring, L., Kleban, M., & Hoffman, C. (2013). Is "walkability" a useful concept for gerontology? *Journal of Housing for the Elderly*, *27*, 241–254.

Ley, D. (1974.) *The black inner city as frontier outpost. The Monograph Series of the Association of American Geographers*. Chicago: Published for the Association of American Geographers by Rand McNally & Co.

Lubrano, A. (2013, November 11). Steep rise in deep poverty Among elderly. *Philadelphia Inquirer*. Available from http://articles.philly.com/2013–10–11/news/42904099_1_deep-poverty-north-philadelphia-law-center

Mayor's Commission on Aging. (2014). *City of Philadelphia: Mayor's Commission on Aging*. Available from http://www.phila.gov/aging/Pages/default.aspx

Norstrand, J., Glicksman, A., Lubben, J., & Kleban, M. (2012). The role of the social environment on physical and mental health of older adults. *Journal of Housing for the Elderly*, *26*(1–3), 121–136.

Ring, L., Glicksman, A., & Kleban, M. (2012, November). *Locating parks, pets & parsnips: Integrating eographic Information Systems (GIS) data into survey research*. Paper presentation at the 65th Annual Scientific Meeting of the Gerontological Society of America, San Diego, CA.

Riordan, K. (2006, November). Philadelphia gears up for an election. *Planning*, 12–15.

Shlay, A. B., & Whitman, G. (2006). Research for democracy: Linking community organizing and research to leverage blight policy. *City & Community*, *5*(2), 153–171.

Technical.ly/Philly. (2013). *The 50 Biggest Employers in Philadelphia and Its Surrounding Counties*. Available from http://technical.ly/philly/2013/02/22/top-50-biggest-employers-in-philadelphia-and-its-surrounding-counties/

ThirdAge, a division of CliftonLarsonAllen. (2012). *Philadelphia corporation for aging: Walkability Impact on Senior Health (WISH) Philadelphia*. Pennsylvania Community Design Charette Report. Plymouth Meeting, PA: ThirdAge.

United States Environmental Protection Agency (U.S. EPA). (2009). *Growing smarter, living healthier: A guide to smart growth and active aging*. Washington, DC: U.S. EPA.

Wang, D., & Glicksman, A. (2013). "Being grounded": Benefits of gardening for older adults in low-income housing. *Journal of Housing for the Elderly*, *27*, 89–104.

Warner, S.B. (1969). *The private city: Philadelphia in three periods of its growth*. Philadelphia PA: University of Pennsylvania Press.
Wieckowski, A. (2010, May). Back to the city. *Harvard Business Review*, 23–25.
World Health Organization. (2006). *Global age-friendly cities*. Available from http://www.who.int/ageing/age_friendly_cities_guide/en/index.html
Young, E. (2001, July). Planning process stimulates Philadelphia neighborhoods. *Planning*, 30.

Part II

Special Topics

9 Commentary: Changing Practice and Policy to Move to Scale

A Framework for Age-Friendly Communities Across the United States

M. Scott Ball and Kathryn Lawler

INTRODUCTION

The realization of age-friendly communities requires as much demolition as creation: undoing decades of public infrastructure investment, reimagining community design and regulation, extracting the one-track inflexibility from current transportation systems, and reconsidering where health care is provided and how it is financed. Communities are faced with decisive demolition because significant public investments were implemented, and continue to be implemented, without consideration for the needs, wants, and desires of changing bodies and minds. It is overwhelming but necessary to face the reality that policies, programs, and funding at all levels of government—federal, state, and local—largely ignored the increasing longevity that resulted from the dramatic medical, scientific, and public health advancements of the twentieth century. As a result of this century-long lapse, physical barriers to successful aging are now deeply embedded in the built environment, and policy barriers are deeply embedded in public institutions.

Whether it was a deep denial of aging, an inability to recognize its consequences, or simple shortsightedness, the resulting health care systems and built environments force many to wage uphill battles to age in place. The challenge is not in the aging, but in the places where it is done (Oliver, 1983). It is the infrastructure, the services, the housing, and the economies of local places that work well or well enough at earlier stages in the lifecycle but that fall down, sometimes to disastrous physical and/or financial consequences, in the later half of life. Meaningful changes to our community places can be made incrementally, but the scale of the increment is large. The paradigms by which we plan streets and transportation systems must be shifted. The paradigms by which we zone community development must be rethought. The focus of medical systems on the delivery of acute care treatments rather than the maintenance of health must be reconceived. Not all need to be shifted at once, but it is counterproductive to make only small changes that simply make dysfunctional paradigms slightly more comfortable.

A growing awareness of the need to plan for an aging society has resulted in seminars, toolkits, panels at numerous conferences, programs, and innovations in service delivery across the country. Architects, planners, gerontologists, homebuilders, trade and professional associations, technology and software designers, and graduate programs in a variety of disciplines have all begun updating their practices to capture a share of the new older-adult market. This good work being done by good people has dramatically increased the awareness of the changing demographics and their potential implications. But this work is almost all at the scale of incremental tweaks and does not yet tackle the difficult work of demolishing the overarching paradigms of land use, transportation, health care, and supportive service delivery. The challenge now is to prepare to implement on a scale that begins to approach the coming need. Exclusive focus on pilots, forums, and toolkit resources is potentially a new form of disengagement, trading the denial of the twentieth century for procrastination in the twenty-first, and could ultimately be as shortsighted as the plans and policies that created the challenge in the first place. These efforts cannot just make the existing paradigms more comfortable; rather, they must instigate the very uncomfortable work of destructive re-creation. This article will briefly review the current state of practice related to the challenges of aging in place and discuss a potential framework for moving to scale. Recognizing that it is not possible to carry out meaningful change at this scale without government, the framework attempts to distinguish the demolition of policies and practices from the demolition of institutions themselves by considering how local work can shape larger policies and practices. Creative disruption is needed—but it is reinvention rather than dismemberment.

THE CHALLENGE

Most post–World War II communities in the United States were built on the premise that development should be segregated by use: regional employment centers from regional residential centers, local shopping areas from residential neighborhoods, high-income housing types from low-income housing types, senior housing from single-family homes, and everything desirable in life from the undesirable, including landfills, heavy industry, and even the location of health care facilities. All of our daily needs have been sorted, segregated, and dispersed across a wide geographic area.

Within these sprawling environments, most transportation infrastructure supports only one way to get around—single occupancy vehicles. Most health care focuses on treating time-limited acute episodes rather than sustaining lifelong wellness. Most housing is segregated by type and use, separating housing for families from housing for those who no longer want a yard, multiple bedrooms, and stairs. For decades, individuals and families of all ages and at all stages in the lifecycle have navigated these environments, spending more and more time in their cars while overall health has declined and health spending has increased. As a larger number of older adults are living longer, their ability to navigate these dispersed and often inadequate or inappropriate living arrangements declines. What was once an inconvenience or problem to work around becomes a deal breaker for the older adult. The inability to change housing type without changing community, to get around when driving is no longer convenient or comfortable, and to maintain health, prevent the onset of chronic disease, and/or manage it should it occur are among the most pressing challenges of adults attempting to age in place. Appropriate housing, transportation alternatives, and quality neighborhoods in which one can maintain an active and healthy lifestyle are as central to promoting healthy aging as traditional services like home-delivered meals, in-home support, and senior centers have been historically.

Recently, public health research has established that it is precisely the division of urban components into segregated zones that contributes to some of our nation's most intractable health problems: declining rates of physical activity resulting from auto-dependent environments and widespread respiratory problems associated with poor air quality (Frank, Franklin, & Jackson, 2004). Mental health and quality of life issues are also profoundly affected by contextual factors that range from the stress and difficulties of lengthy work commutes to isolation-related depression for those retired or unable to maintain employment. Poor and minority communities are especially vulnerable, based on their disproportionate exposure to numerous environmental pollutants, crime, and poor access to health care. With a mounting body of evidence documenting the relationships between sprawling development patterns and negative health outcomes, as well as the relationships between compact pedestrian-oriented patterns and positive health outcomes, it seems clear that, for the future to be any different from the present, the planning and zoning regulations that create unhealthy sprawling urban conditions and prohibit healthy, compact, walkable communities must be redone.

When compared, the resources in the medical system dwarf the occasional afterthought given to sidewalks, parks, and civic spaces, although the latter are tangible investments in the places where healthy routines can be reinforced in a continuous and passive manner. Neighborhood resources are unfortunately valued as amenities only, and although comparatively cheap across their useful existence, they are viewed as luxuries rather than long-term health investments. Health care facilities continue to improve and grow, even in a significant economic downturn, while public space amenities are the first programs to be cut when budgets tighten. All the while, national, state, and local transportation dollars are almost exclusively invested to improve the ability of drivers to drive, neglecting the growing portion of the population of non-drivers. This disparity, on the one hand, highlights the lack of awareness of the ways in which the built environment is intrinsically tied to lifelong well-being and, on the other hand, illustrates the lack of administrative systems capable of assigning value to the health benefits of environmental and infrastructure planning and decision making.

U.S. Census demographic projections indicate that the status quo since World War II is unsustainable:

- There will soon be more than twice as many older adults than there ever have been.
- Between 2005 and 2030, the population of people older than 55 will double across the nation.
- The populations of other age groups will not grow nearly as much.
- Today, 1 out of every 10 persons is an older adult. By 2030, this will be 1 out of every 5 (Centers for Disease Control, 2007).
- The percentage of older adults who are very old will be much larger and in many ways different from the generation before.
- More people will live to be older than 85, and it will be increasingly common to live more than 100 years (U.S. Census Bureau, 2010).
- Much of the older adult population will be single or living in multi-generational households, creating a greater demand for more diverse housing options in neighborhoods: smaller housing units, cohousing, shared housing, accessory units, apartments, and other housing arrangements that make it possible to stay in the same community as life stages and circumstances change.

    o Currently, 40% of women and 20% of men older than 65 years of age live alone, but this is a generation in which less than 10% of marriages ended in divorce. With more than 20% of baby boomer marriages ending in divorce, the majority of the boomer population may be single as older adults (U.S. Census Bureau, 2010). At the same time, while much of the older adult population will be single, men are living longer, and more and more families are living in multigenerational housing.

- Older adults typically stay in or close by the homes and communities they know.

    o Eighty percent of all residents of all senior housing facility types live less than 25 miles from their prior homes (American Seniors Housing Association, 2009).
    o Ninety percent of those 65 and older surveyed by AARP responded positively to the statement: "I would like to be living in the same local community 5 years from now" (AARP, 2005).

- The majority of the baby boomers live in use-segregated, suburban areas. Seventy percent of baby boomers lived in the suburbs and accounted for roughly 31% of the total suburban population in 2000 (Frey, 2003).

- Fewer workers will be supporting more non-workers. The dependency ratio—the proportion of working-age populations (aged 15 to 64) compared to nonworking-age populations (aged 0 to 15 and older than 64)—will shift. There will be more people demanding services and fewer people available to deliver these services (U.S. Office of Personnel Management, 2008).

These demographic shifts will exacerbate a range of challenges that did not have to exist. The task now is to institute the routine repair of the sprawling built environment and the routine reorganization of transportation and support services so that they can facilitate healthy living in the community. It should already be commonplace and unexceptional to require that new development support the health of the community and accommodate the needs of people across their lifespans. In addition, and at a very minimum, moving forward, no public money should be directly or indirectly spent in a way that assumes eternal youth.

THE STATE OF CURRENT PRACTICE

There has been a ground swell of interest and work on making communities more age-friendly since 2000. Often it has drawn on models like the NORC (Naturally Occurring Retirement Communities), Supportive Services Program, and other initiatives that trace their genesis to the mid-1980s. Extensive research has already contemplated what communities need to do to accommodate their growing older adult populations. This has included highly detailed recommendations on community infrastructure design (sidewalks, curb cuts, crosswalks, wayfinding, street-sign design, etc.), recommendations on housing-type mix and location, very sophisticated finance reforms, and reimagining entire long-term service and support systems. While there will always be more to learn, what has been researched and catalogued far outstrips standard practices. At the same time, there are encouraging emerging practices operating outside of the mainstream. Communities like Atlanta, Georgia; New Orleans, Louisiana; Philadelphia, Pennsylvania; New York, New York; and many others across the country have begun programs and initiatives to make their communities better suited to meet the needs of people of all ages and abilities. This extraordinary work has bridged multiple disciplines and piloted innovations in land use, transportation, housing type and location, senior-center programming and design, and community-based service delivery.

Despite the good work to date, these communities are the outliers, and almost all older adults are living and will continue to live in communities that make it difficult to age successfully. New development predominantly follows the old patterns that we know to be problematic. The innovative work is fragile and pioneered by courageous and determined individuals who have found ways to break down silos and to work across departments, disciplines, and funding streams. These programs have shown how transportation can be more flexible to suit the non-driving non-commuter; how housing can be combined with services; how accessible, affordable, and appropriate housing for people of all ages and abilities can be located within existing neighborhoods; and how communities and services can meet people where they are, rather than force individuals to limited and outdated choices of services and supports. But all too often they rely on waivers, exemptions, and the cooperation of officials who can either be reappointed elsewhere or lose an election in the next cycle. Almost all depend on a creative mixture of funding streams that combine local, philanthropic, and occasionally state or federal dollars to deliver services or change policies. Moving against the grain of entrenched standard practices requires heavy and consistent labor. But none of these programs has dedicated ongoing funding support to undo the decades of investments, policies, and regulations that have made aging in the community such a challenge.

Age-friendly communities work cannot survive as a stand-alone program or pilot effort. It cannot be framed or funded as a program du jour, the latest and greatest in aging services, or planning paradigms, or it will eventually lose momentum (and funding) when replaced by the next catchphrase or fad. Rather, this work is dependent on large-scale reform, in need of the kind of metamorphosis that, as an example, the Internet has catalyzed in every aspect of society in a very short time. Not long ago, most businesses could not imagine how they might function without typing pools, hand-drawn graphics, and slide rules. Technology has been and continues to be a disruptive and powerful change agent in every system and process of every business, government, and almost every household. It affects everything and everyone. Technology is a new and permanent state—much as longevity appears to be a new and permanent state. The task is to adjust current systems of service, funding, regulation, and policy to reflect this new state and to enable the wealth of evidence and experience in hand to disrupt and reinvent systems from within. Local projects and pilot initiatives alone simply cannot and will not work; they must instigate systems change if they are going to create full-scale disruptive change needed to produce lifelong communities.

A FRAMEWORK FOR CHANGE: ACT LOCALLY, THINK NATIONALLY

If age-friendly communities are to emerge at any scale from innovative local efforts, then those local efforts must be more intentionally targeted toward reshaping federal institutional policies, practices, and funding streams.

Without a strategic framework that provides broader context for local initiatives and helps to clear the way at the federal and state levels, pilot programs will rely too heavily on the ingenuity, energy, and bureaucratic acrobatics of highly motivated individuals and organizations willing to swim upstream against policies that do not provide the options people need throughout their lifetimes.

In the ideal framework, pilot programs should function as beta tests at the local level, the results of which are sent to the federal level, where they are used to adjust national policy to the evolving needs of individuals and neighborhoods. Funding should forecast and anticipate the need for substantial reforms and fund "foot-soldier" pilot programs as a way of introducing oncoming changes in federal intent. Pilot programs would not be the end goal—a kind of evidence that things are working—rather, they would be a way of acknowledging, reforming, and realigning the things not working in national efforts. If the goal is to improve the lives of many older adults, local pilot programs should be evaluated on their success in adjusting larger programs and systems, not solely on their success in improving local conditions. As we pilot efforts to address these deficiencies, we should celebrate accomplishments without claiming success. Success is achieved when the paradigms that produce the unlivable conditions are changed, not when relief is provided for a few.

Undoing or Redoing Key Policies Through Local Work

Deficiencies in federal legislative initiatives, funding, and regulations have a tremendous impact on the ability of communities to support aging populations. Within the context of these large-scale deficiencies, local pilot programs can become intentional and targeted components of the necessary larger solutions. Federal agencies could learn to glean from successful pilots and make adjustments accordingly. Pilot programs should be able to articulate which federal policies have hampered their success and which have required scarce, private-foundation funding to close the gaps. Without addressing the larger concerns, pilot programs cannot "pilot" widespread replication, but will instead remain one-off successes that might serve only to mask larger deficiencies. Local efforts must target their efforts towards successful systemic change and national efforts must learn to integrate pilots into market testing and strategic adaption processes. Some examples are detailed below.

Americans With Disabilities Act

The ADA of 1990 (42 U.S. Code Chapter 126) opens with findings made by Congress that establish the need for the legislation. The very first of these findings references the aging population: "Some 43,000,000 Americans have one or more physical or mental disabilities, and this number is increasing as the population as a whole is growing older" (Americans with Disabilities Act, 42 U.S. Code Chapter 126, 12101. (a) (1)). After this first mention in the legislation, very little of the ADA's evolution and implementation has gone on to address the unique needs of aging populations. As projected, the population has aged, and according to the U.S. Bureau of the Census, Survey of Income and Program Participation, 52% of the population older than 65 is disabled compared to 16% of the population younger than 65. Disability is much more of an issue for older adults than for any other segment of the population.

Local- and state-level age-friendly community programs could be an important part of an effort to address deficiencies in the ADA at the federal level, where the ADA is administered and enforced as a civil right, and at the state level, where ADA policies are statutorily adopted. Pilots in community accessibility could be integrated into the processes through which federal ADA guidelines are reviewed, updated, and implemented (Ball, 2012). The lessons learned at the local level about barriers to aging in place and ways of overcoming them might be better institutionalized and broadcast if done in conjunction with the United States Access Board and integrated into the American National Standards Institute's processes for developing accessibility standards (ANSI 117.1). In some respects, age friendliness may be more than just an admirable community attribute; a close reading of the ADA would suggest that many forms of "age unfriendliness" could be considered discriminatory.

Older Americans Act

The Older Americans Act (OAA) of 1965 is the first federal attempt to coordinate a comprehensive set of services to support older adults. This act created a multitiered National Aging Network comprising the Administration on Aging at the national level, State Units on Aging, and Area Agencies on Aging serving every community in the nation. The sweep of the OAA mandate staggeringly overshoots the funding historically provided: to ensure equal opportunity to the fair and free enjoyment of adequate income in retirement; the best possible physical and mental health services without regard to economic status; suitable housing; restorative and long-term care; opportunity for employment; retirement in health, honor, and dignity; civic, cultural, educational, and recreational participation and contribution; efficient community services; immediate benefit from proven research knowledge; freedom, independence, and the exercise of self-determination; and protection against abuse, neglect, and exploitation. Over time, the programs and services have become heavily prescribed and mostly evaluate the number of items or services delivered, rather than the outcomes achieved. As a result, in many communities, OAA services and the network that delivers them have narrowed, having been shaped primarily by the funding allocated to the Titles of the Act rather than to the older adults they are committed to serving. It is time to modernize these services to suit the needs of modern older adults, and some local communities are already leading the way. Much of the existing age-friendly community work could inform the next reauthorization of the OAA if the pilot projects were designed from the outset to create systems-level changes. This local work should be designed to demand greater flexibility at the federal level, allowing programs to serve *people*—not just a specified quantity of meals, rides, or hours of care. Area Agencies on Aging provide every community in the United States a network and, in many places, a possible institutional home for innovation in service delivery and translation back up to the federal level.

National Standards for Healthy Planning and Zoning

Local planning is enabled by federal law, but it is not guided by federally legislated standards. A national mandate allows states and localities to undertake "planning and zoning." In the 1926 landmark Supreme Court case of Ambler v. Euclid, the justices declared that state and local governments have the authority to promote "the health and safety of the community" by planning and zoning, but it is left entirely up to those localities to determine what exactly planning and zoning are. The Euclid case provided no guidance or standards for what constitutes healthy planning and zoning, and legislative efforts to enact standards simultaneous to the Euclid legal case (the Standard State Zoning and Planning Acts of 1926 and 1928) failed to be adopted by Congress.

Recent research has clarified the relationship between urban form and health (Frank et al., 2004). It is now known that healthy land use policy must allow for, rather than prevent, integrating uses so that jobs, services, and retail are within walking distance of housing. Given the degree to which local planning and zoning efforts have pushed these uses apart and has thus diverged from healthy practices over the past century, it is time to raise again the possibility of national standards for planning and zoning, similar to the efforts in the late 1920s to enact federal Standard State Planning and Zoning Acts. Zoning authority should be evaluated on whether it has been exercised in a manner that is healthy for the community, particularly for vulnerable populations such as older adults.

Federal Tax Policy Related to Real Estate Investment

Tax policy has a significant influence on the housing options available in a community. Numerous Internal Revenue Service policies currently give preferential treatment to investments that work against the needs of an aging population. Probably the most significant example is the mortgage interest tax deduction that allows owners to write off the interest paid to banks for loans on first and second homes (Dolbeare, 1985). These policies favor not only ownership over renting, but large and multiple homes over the smaller holdings that are more manageable for older adults. Within the categories of housing specifically for seniors, the IRS recognizes independent living types as "passive investments" eligible for Real Estate Investment Trust (REIT) status, while housing types that integrate any level of support services are considered "active investments" not eligible for REIT status. The result of this tax advantage will be that more independent senior housing will be built in larger quantity than assisted living or skilled nursing housing that meets very real needs.

Section 8 Housing Vouchers and Medicaid Waivers

Long-term care by definition requires both housing and health care services (Lawler, 2001). Nursing homes have long combined the two into one model. But for any other, less intensive setting, it is extremely difficult to combine both health services and housing for low-income older adults in need of support. They or the development that intends to house and support them must obtain both Section 8 vouchers and Medicaid waivers, but the process to apply, qualify, and receive assistance through these two programs has almost nothing in common except extremely long waiting lists. As a result, all too often, an individual has housing but no services or services but no housing. Either circumstance can force individuals unnecessarily into institutional settings that readily provide both.

States can set eligibility policies that enable the coordination of housing and supportive services, very specifically Section 8 vouchers and Medicaid waivers. Recent examples in the Money Follows the Person program have demonstrated how the waiver and voucher can align at the right time, allowing low-income individuals to leave institutions and live in the community. States working under the Balancing Incentives Payment Program, as part of a series of changes to increase the number of people receiving long-term care in the community and decrease those in institutional settings, are expanding waivers and providing unbiased, conflict-free counseling to empower individuals and families to make the best decisions about their support. This alignment of housing vouchers and Medicaid waivers in this program could be expanded to make it possible for many older adults to find the right combination of housing and support they need to remain in the community, the ultimate goal of age-friendly work.

Reinventing What Gets Measured to Achieve Results Locally and Nationally

What is often missing from the story and evaluation of local accomplishments is the most important piece: the quality of coordination among various government agencies. For example, the best transportation program will quickly fall short if it is organized in a community with inadequate housing (wrong type of housing, wrong price of housing, or both). The best housing and transportation systems will not create an age-friendly environment without a flexible, community-based, long-term care system. The interactions matter because, whether good, poor, or nonexistent, they are what make it possible for people of all ages and all abilities to live in a community throughout their lifetimes. Pilot programs are too often judged by their immediate results in creating better living circumstances for a specific person or group of people in a neighborhood, when the more significant impacts to institutional processes are those that will not reveal themselves until the next 1,000 projects are implemented elsewhere. Institutional changes, particularly interdepartmental changes, are far more likely to "move the needle" on the issue that determines a community's livability and should be evaluated, measured, and celebrated relative to immediate results.

What is measured, however, often determines what and how things get done. Pilot programs should explore and test new metrics to guide the award and administration of federal and state funding. No matter the program's intention or mission, it is the evaluation, the payment structure, the annual report, or the funding formula that will determine what gets done. Some examples: As long as programs count and pay for the number of meals provided in a congregate setting, aging services providers will continue to be forced to drive people to congregate meal facilities, feed them, and keep them there for a set number of hours, whether the individual finds this beneficial or not. Meanwhile, the congregate meal program was designed originally to battle isolation, with the meal serving as the means rather than the end in itself. But socialization and the outcome of meals provision on overall health are not measured. The number of meals is. If a program will pay for a day of adult day care provided that there is a minimum of six hours of care, then only six hours of care will be delivered. The hours are measured, as opposed to the improved health of the older adult or the caregiver.

The recent two-year reauthorization of the federal transportation bill MAP 21 (Moving Ahead for Progress in the 21st Century) has begun to incorporate performance measures in its funding criteria (Federal Highway Administration, 2012). For some aspects of the new bill, it is not just about the number of miles of road or bridges constructed, but how well that infrastructure performs for the people who use it. It is just a start, but certainly the kind of example in which altering what gets measured can begin to alter what gets produced. Peer review processes could play a valuable role in evaluation. Rather than establish a dialectical relationship between funder and funded, federal performance metrics might incorporate peer reviews between funded projects as a means of ensuring that the highest standards, rather than minimum standards, guide development.

Local pilot programs can help define the types of performances that federal and state agencies seek to underwrite; working back from those achievements, funding and award metrics can be reverse engineered to achieve the desired results. At all levels—local, state, and federal—longevity should prod a rethinking of what gets measured and what outcomes are desirable at different stages in the lifecycle.

CONCLUSIONS

When the people lead, the leaders will follow. In a political environment that frustrates both ends of the political spectrum and every person and idea in between, it is quite appealing and possibly the only option to contemplate change from the ground up. But more than gridlock or a dearth of new ideas at the national level, change that has been driven by the local level is far more likely to ensure that what is supplied (policies, programs, and regulations) meets the demands of an aging society that has never aged at this scale before. The United States and most developed countries around the globe find themselves in a period of significant reinvention. Communities were built and massive programs were designed without consideration to the longevity that science, medicine, and public health worked so hard to achieve in the last century. As American communities seek to redo and to reimagine how they will function to accommodate a longevity that society hopes will continue, it is essential that policies and programs reflect as closely as possible the realities of older adults and their daily experiences. These adjustments can best be made when local action drives broader change with strategic acts of systemic reinvention, such as those suggested in the Framework for Change section in this paper.

Yet, driving broader change is more than most age-friendly community programs have achieved to date. The argument made in these pages will require a different approach on both sides; all involved must change what they are doing now because current practice is never going to meet the needs. National and state policy must accept the lessons of local communities, but local work must be designed from the start and implemented throughout as systems change, not just provide short-term and highly localized benefits.

Current efforts must not rest on the satisfaction of improving a single community. Without strategic efforts aimed at long-term change, local efforts can quickly become consumed in human interest news stories, best practice reports, and highly publicized ribbon cuttings that mask the need for a massive overhaul of the fundamental deficiencies in federal institutional implementation of programs critical to older adults. To avoid this fate, the current practice of age-friendly community work must engage in creative destruction: challenging entrenched interests, stagnant bureaucracies, and obsolete programs.

This will not be easy or any less complicated, but the demographics demand more than the field is supplying today. The faster the work can be brought to scale, the more significant and meaningful the difference to future and, perhaps, even current older adults.

REFERENCES

AARP. (2005). *Beyond 50.05: A report to the nation on livable communities: Creating environments for successful aging*. Washington, DC: AARP.

American Seniors Housing Association. (2009). *The independent living report*. Washington, DC: American Seniors Housing Association.

Ball, M. (2012). *Livable communities for aging populations: Urban design for longevity*. Boston, MA: John Wiley & Sons, Inc.

Centers for Disease Control. (2007). *State of aging in America*. Atlanta, GA: Centers for Disease Control.

Dolbeare, C. (1985). *Federal housing assistance: Who needs it? Who gets it?* Washington, DC: National League of Cities.

Federal Highway Administration. (2012). *Moving ahead for progress in the 21st century act*. Washington, DC: Federal Highway Administration.

Frank, L., Frumkin, H., & Jackson, R. (2004). *Urban sprawl and public health: Designing, planning, and building for healthy communities*. Washington, DC: Island Press.

Frey, W. (2003). *Boomers and seniors in the suburbs: Aging patterns in US Census 2000*. Washington, DC: Brookings Institute.

Lawler, K. (2001). *Aging in place: Coordinating housing and health care provision for America's growing elderly population*. Cambridge, MA: Joint Center for Housing Studies of Harvard University.

Oliver, M. (1983). *Social work with disabled people*. New York, NY: McMillian Publishers.

U.S. Census Bureau. (2010). *Current population survey, annual social and economic supplement*. Washington, DC: U.S. Census Bureau.

U.S. Office of Personnel Management. (2008). *An analysis of federal retirement data*. Washington, DC: U.S. Office of Personnel Management.

10 Transforming the Way We Live Together

A Model to Move Communities from Policy to Implementation

Laura Keyes, Deborah R. Phillips, Evelina Sterling, Tyrone Manegdeg, Maureen Kelly, Grace Trimble, and Cheryl Mayerik

INTRODUCTION

For the first time in history, communities across the United States must consciously prepare for the forecasted doubling of the number of people older than 65, from 49 million to a projected 89 million people by the year 2050 (Jacobsen, Kent, Lee, & Mather, 2011). As these numbers increase, government services such as meal provisions, senior center programming, and on-demand transportation for medical visits, which have traditionally helped the country's oldest and most frail population segment, will not be enough to support the demand of this growing older adult population. Instead, a focus on community planning and development designed to keep people independent and healthy in their communities longer will be required. These preparations include ensuring adequate housing, expanding opportunities for new mobility choices for non-drivers, improving access to basic services, and creating opportunities for healthy living.

Despite this growth among older adults, a survey conducted by the National Association of Area Agencies on Aging of 1,500 communities revealed that very few communities across the nation had undertaken a comprehensive assessment of livable communities for all ages. Although communities indicated advances in public safety, adult education, and community volunteerism, many expressed that the lack of resources and services made it difficult to address the three core needs of older adults: transportation, housing, and financial security (National Association of Area Agencies on Aging, 2011). Significant challenges exist to aging in place, including the lack of community-based services and infrastructure to keep people healthy and independent for as long as possible (Knickman & Snell, 2002), and, all too often, individuals find that the community in which they have lived for years no longer meets their needs as they age, specifically with regard to mobility options and adequate and affordable housing. A community assessment assists residents and local governments with the identification of a community's assets, or lack of assets, that either support or hinder aging in place.

The Atlanta region has one of the country's fastest growing older adult populations, expected to increase from 1 out of 10 residents older than 65 in the year 2010 to 1 out of 5 by 2030. In an effort to support its mission for healthy and independent aging, the Atlanta Regional Commission (ARC) is working with communities throughout its 10-county region to create places in which people of all ages and abilities can live as long as they would like. ARC's Lifelong Communities Initiative is a community planning and public health response to ensure that communities have an array of housing types that appeal to individuals both young and old, opportunities for healthy living with ways to get around that meet the needs of individuals who do not drive, safe sidewalks and interesting places to walk, and convenient access to shopping and basic services. ARC created its Lifelong Communities Initiative to show that age-friendly communities and livable communities are related. The following review of literature and resources discusses the necessary elements of a Lifelong Community and the important role these elements have in supporting the three goals of a Lifelong Community: provide housing and transportation options, encourage healthy lifestyles, and expand access to services. The chapter will further demonstrate ARC's efforts to help communities advance the Lifelong Communities Initiative from planning to implementation to help residents more fully realize how mobility options, affordable and accessible housing, and access to services can truly help people stay in their communities as they age through its Lifelong Communities Case Study in the community of Mableton.

VALIDATION OF THE ELEMENTS FOR A LIFELONG COMMUNITY

In developing a strategic approach to creating its Lifelong Communities Initiative, ARC assessed the necessary elements a community must have to create a Lifelong Community. The elements include connectivity, pedestrian access and transit, neighborhood retail and services, diversity of dwelling types, healthy living, and social interaction. These particular elements are important to ARC's Lifelong Communities model because the initiative's purposeful effort is to integrate the needs of an aging population into all levels of service, health, and infrastructure and to develop places that are fully accessible from inside the dwelling to down the street and into the doctor's office, restaurant, park, or store.

Connectivity between housing and basic services and the pedestrian infrastructure to access the community offer mobility options to older adults, especially when they may no longer be able to drive. Seniors' external environment is an integral part of their quality of life (Neal et al., 2006; Masotti, Fick, Johnson-Masotti, & Macleod, 2006; Tinker, 1997; World Health Organization, 2007). Connectivity or "easy access, either in the neighborhood or via public transportation, to shopping, cultural and educational opportunities, health providers, and other service providers" contributes positively to seniors' perception of a good neighborhood and their ability to age in place (Philadelphia Corporation for Aging, 2008). More than 80% of people 45 and older prefer to have services close to where they live, with the majority preferring to live near medical facilities, shopping centers, grocery stores, and places of worship (AARP, 2003). Through a well-connected transportation infrastructure, older adults may have access to more mobility options when they can no longer drive. Pedestrian access and transit affect the mobility of seniors, increasing their independence and easing access to required services, particularly medical appointments (Philadelphia Corporation for Aging, 2004). A report on findings revealed that pavement conditions, pedestrian safety, and access to transit, public spaces, and green areas can also improve seniors' social interaction, overall mobility, and quality of life (New York Academy of Medicine, 2012; Philadelphia Corporation for Aging, 2006; Portland State University, 2009).

Clustered retail services that are within walking distance to where older residents live alleviate the need for transportation support, enable older residents to participate in their community, and help them deal with their increasing loss of functional capacity as they get older (Richard, Gauvin, Gosselin, & LaForest, 2008; WHO, 2007). Age-friendly business practices, such as senior discounts and separate queue lines, positively impact seniors and provide economic benefits to the surrounding neighborhood (New York Academy of Medicine, 2012; WHO, 2007). Accessibility to relevant services near senior communities is also recommended for creating age-friendly neighborhoods (King & Farmer, 2009; Masotti et al., 2006; Philadelphia Corporation for Aging, 2006).

Lifelong Communities offer multiple housing options at various price points to allow people to remain in their communities as their needs change. Because older adults spend a good deal of time at home, housing plays an integral role in quality of life and health (Tinker, 1997). This is especially critical since 90% of older adults want to age in place (AARP Public Policy Institute, 2011); one study found that almost half want to remain on the same block in their neighborhood (Philadelphia Corporation for Aging, 2008). Aging in place requires dwelling units and neighborhoods to be flexible and adaptable to seniors' needs and to accommodate a variety of lifestyles (Ziebarth, Erm, & Hwang, 2010).

Comprehensive planning efforts, including zoning and land use policies, need to support residents of all ages and move toward having "lifecycle communities" in which residents are more able to age in place (Howe, 2001). Howe (2001) suggested looking into different housing alternatives, such as home sharing, accessory apartments, and Elder Cottage Housing

Opportunity, commonly known as ECHO Housing. Growing in popularity, ECHOs consist of portable cottages placed on single-family house lots to provide accommodations for older adults. It is important to emphasize the need to incorporate features benefiting both young and old residents, preventing age segregation in neighborhoods (Ziebarth et al., 2010). Diversity in the housing stock offers people the opportunity to make choices about their future housing based on their current needs and lifestyle.

Communities that encourage and support healthy aging also provide access to activity, healthy foods, and health care. Facilitating healthy living among older adults is important for communities that want to encourage seniors to age in place (WHO, 2007). Plouffe (2011) asserted that age-friendly communities feature accessible physical and social environments promoting active aging, socialization, and healthy lifestyles among older adults. Resting areas, green spaces, clean and safe walkways, and proximity to senior-friendly retail services all contribute to an age-friendly community and to how older adults maintain their health (Jacobs as cited in New York Academy of Medicine, 2008; Masotti et al., 2006; MetLife Mature Market Institute, 2010; WHO, 2007). Social interaction positively affects the ability to age in place, as well as overall health and wellness among seniors (Masotti et al., 2006; Richard et al., 2008). Communities offering diverse cultural and recreational activities for seniors help improve older adults' health while raising the community's overall level of engagement and economic competitiveness (New York Academy of Medicine, 2012; MetLife Mature Market Institute, 2010; Philadelphia Corporation for Aging, 2006; WHO, 2007).

MOVING COMMUNITIES FROM CONCEPT TO IMPLEMENTATION

The Atlanta region, with its rapid growth of the 65-plus population, offers an opportunity to apply the theoretical concepts for developing Lifelong Communities. ARC is the planning, development, and intergovernmental coordination agency for the 10-county region surrounding Atlanta, Georgia, in addition to serving as the region's Area Agency on Aging focused on planning, funding, and administering programs and services for older adults. ARC collaborates with diverse public and private health and social service agencies, hospitals, and planning departments at the county and city levels to promote the well-being and quality of life of older adults. Although basic services like home-delivered meals and personal support will continue to be the foundation of the Area Agency on Aging's work, the services are simply not enough to meet the growing demand on the part of older adults and caregivers for these services.

In 2007, ARC conducted a survey of older adults in Metro Atlanta aged 55 and older to better understand the characteristics and preferences of the

Atlanta region's future older adult population.[1] These data offered local communities a glimpse of the needs of a rapidly growing population and provided opportunities to discuss the expectations of this generation for housing, transportation, and services, as well as the traditional funding sources for aging services (Atlanta Regional Commission, 2007). Key findings indicated that a majority of older adults think that the Atlanta region is a good place to retire. Most of the respondents said they had lived in the Atlanta region for more than three decades, and 64% planned to remain in their homes or communities. Of the major categories of health, housing, long-term care, civic engagement, and transportation, most concerns centered on transportation. Although more than 88% of respondents indicated that they currently owned and drove their own vehicle, 15% were already having trouble getting where they needed to go. When asked what they would do when they could no longer drive, more than 57% said they would be driven around by other people; however, this percentage of respondents also stated that they did not know who the others were who would drive them. Access to reliable transportation plays a key role in supporting the concept of aging in place. The results of this survey bolstered ARC's efforts to create its Lifelong Communities Initiative, a community guide to develop responses to the needs of older adults for housing, transportation, healthy lifestyles, and access to necessary services (Figure 10.1). The goals of the Lifelong Communities Initiative are met by a community's ability to address new housing options, alternatives to the car, walkable communities, physical and social well-being, preventative health care options, and expansion of available health and supportive services.

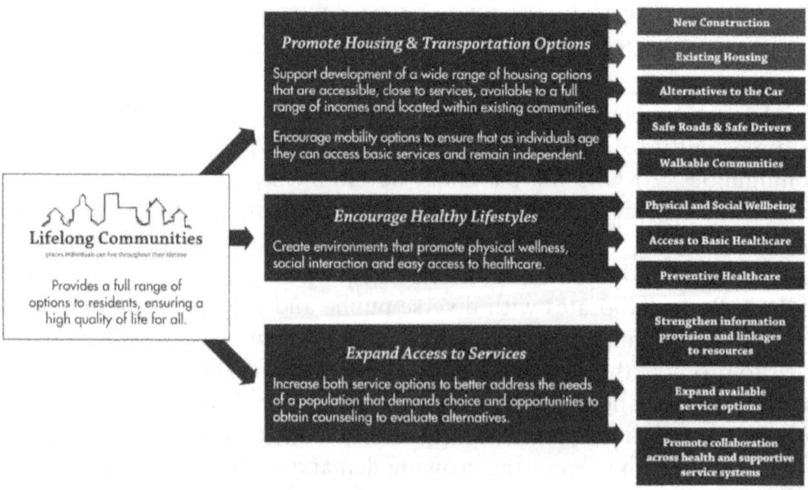

Figure 10.1 Lifelong Communities Initiative

Following the survey, ARC staff met with elected officials, county staff, and community stakeholders in each county within ARC's 10-county region between 2007 and 2008 to discuss and assess the needs of older adults living in their jurisdictions. This effort was important because it allowed ARC to work with each county to determine the feasibility of implementing the Lifelong Communities Initiative and whether an implementation effort designed around the three Lifelong Communities goals would meet the needs of the county's residents. One such effort included the development and passage of a senior housing ordinance in Woodstock, Georgia. This ordinance linked new senior housing with requirements for pedestrian facilities and accessibility. Two years after the adoption of the ordinance, the city's first affordable senior rental housing within walking distance to basic services and resources was established. ARC also created the first of many Lifelong Communities planning tools with the "aging in place tool kit." This tool kit includes easy-to-understand language, explaining the connections between housing, zoning, transportation, development, and the needs of an aging population (Ball, 2008). With early support from elected officials and effective messaging, ARC garnered the necessary support for a Lifelong Communities charrette, a nine-day planning and design workshop that examined specifics steps communities needed to create a Lifelong Community.

Later in 2009, ARC partnered with Duany Plater-Zyberk & Company, an internationally known architecture firm specializing in traditional neighborhood development and relationships between community design and aging in place, to host a region-wide planning effort with over 1,500 diverse professionals. Over the course of nine days, professionals from a variety of fields, including transportation, planning, health, architecture, geriatrics, housing, and mental health came together to apply the Lifelong Communities Initiative to six Atlanta communities that were selected through a competitive process and reflected demonstrated commitment. The goal was not only to develop conceptual master plans for each community, but also to establish a set of principles and guidelines for further development based on the initiative.

ARC recognizes that a single "best model" of a Lifelong Community does not exist in the Atlanta region. Rather, ARC works with communities in a variety of capacities to apply the Lifelong Communities Initiative and Principles. Still, the current zoning codes and transportation policies make it difficult to implement the necessary Lifelong Communities Principles uniformly in many of the region's jurisdictions. However, increasing numbers of local jurisdictions are becoming new partner communities by demonstrating leadership and devoting new resources to support programs. Subsequently, ARC integrated the Lifelong Communities Initiative into the regional land use and transportation plan, Plan 2040, and its planning grant program to local communities, the Livable Communities Initiative.

PUTTING THE LIFELONG COMMUNITIES INITIATIVE INTO ACTION: A CASE STUDY IN MABLETON, GEORGIA

Following the Lifelong Communities charrette, ARC received funding from the Community Innovations for Aging in Place Grant program of the United States Administration on Aging to support the implementation of the Lifelong Communities Initiative in Mableton: Lifelong Mableton. ARC selected the area of Mableton in Cobb County, Georgia, to test the full implementation of the Lifelong Communities Initiative. This allowed ARC to describe the effectiveness of the Lifelong Communities Initiative's tools and understand how better to mobilize communities from planning and start-up to implementation and action.

The Lifelong Mableton Project encompasses the unincorporated area of Mableton and the southernmost unincorporated part of Cobb County, Georgia. This area includes a very diverse and low-income population and simultaneously a high percentage of older individuals (up to 60% of the population in some locations). Both African American and Hispanic populations are significantly represented as well (U.S. Census, 2008). ARC selected Mableton because, over the past 10 years, it has been a focal point for community improvement in the region. Lifelong Mableton is a project focused on building capacity—through planning, design, programming, and community involvement—to transform Mableton into a livable community for individuals of all ages. With the participation of the community and local planning staff, ARC created a work plan for Mableton based on achieving the Lifelong Communities goals for housing, transportation, healthy living, and access to services. ARC also conducted a comprehensive evaluation plan, which encouraged reflection and necessary adjustments to work efforts and programs at each milestone to ensure success. Through ARC's formal evaluation process of the Lifelong Mableton Project, four themes emerged as necessary for creating a sustainable approach to the development of a lifelong community: innovative partnerships through organizational collaboration, integrating Lifelong Communities principles into local planning, community engagement through citizen participation, and addressing the needs of underserved populations.

Theme 1: Innovative Partnerships Through Organizational Collaboration

Early messaging and aligning the vision of Lifelong Mableton with local community priorities was important for gaining local trust and participation in the process. Finding common themes and opportunities to work with other strategic alliances proved beneficial. For example, ARC partnered with Cobb County's Board of Public Health to find a common goal to improve sidewalk access to the local middle school. The Board of Health championed a Safe Routes to Schools project at the local middle school, which increased opportunities for walkability for the schoolchildren and led to a new stoplight and intersection improvements at a nearby location used by older adult walkers.

Recognizing the need for continued organizational collaboration with the Lifelong Communities Initiative, ARC quickly strengthened community partnerships, including the establishment of the Lifelong Mableton Partnership Team, a consortium of business leaders, civic leaders, and government officials, as well as the Lifelong Mableton Mental Health Collaborative. Although not through a formal agreement, the partnership met regularly to discuss the progress of the Lifelong Mableton activities and to determine their potential roles. These collaborative efforts allowed partners to advance initiatives more quickly, share resources, and leverage funding. ARC also cultivated an interest in the well-being of older adults within Mableton's established organizations. Through their new understanding of the principles required to build a lifelong community, these organizations will continue to support the Lifelong Communities Initiative long after ARC completes the Lifelong Mableton Project.

Theme 2: Integrating Lifelong Communities Principles Into Local Planning

To focus on redevelopment and to remove policies creating barriers to housing options and pedestrian access, Cobb County government developed a new zoning effort called the Mableton Form-Based Redevelopment District. The Mableton Form-Based Redevelopment District integrated the three Lifelong Communities' goals for mobility and housing options, healthy lifestyles, and access to services into the local zoning code to incentivize the development of a range of housing, new retail, and pedestrian-friendly development. The code supports the vision of Lifelong Mableton by allowing for a town square[2] and gathering place for residents. Further, it proposes a redesign of a major roadway through the Mableton area to include a pedestrian and bike path connection to the grocery store and other retail amenities, as well as to the Silver Comet Trail, the longest walking and bike path in Georgia. Furthermore, the code proposes housing of differing sizes, styles, and price points to give longtime residents of Mableton housing options that may allow them to stay in their neighborhoods as their needs change. As a result of the initial planning effort, the school district redesigned the school to incorporate better the Lifelong Communities principles and align with the vision for a future town square, as planned in the Lifelong Mableton Redevelopment District. The newly constructed school opened in the fall of 2013.

The new code allows for a mixture of single-family and multifamily housing so that older adults will have a choice of housing options when choosing to downsize; it also allows for increased density, along with a mixture of commercial and retail uses with a focus on pedestrian access (CIAIP Technical Assistance Team, Center for Home Care Policy & Research Visiting Nurse Service of New York, 2012). The area of Mableton subject to the new code includes the same two-mile radius that encompasses the Lifelong Mableton Project study area. To move from policy to implementation, community leaders created redevelopment plans attractive to developers, aligned the code with the community's vision to create a Lifelong Community, and legalized them through the county's zoning ordinance.

Theme 3: Community Engagement Through Citizen Participation

Recognizing the need for robust civic input and engagement, ARC created a Lifelong Communities Citizen Assessment Survey. ARC administered this survey in Mableton to collect more input from residents. While collected after ARC had already initiated work efforts such as the Mableton Form-Based Redevelopment District, the Mableton survey results confirmed that Mableton residents are most concerned about improving social settings and access to services. More specifically, they expressed a desire for more active and passive open spaces and more places to gather in light of the fact that Mableton lacked a formal town center. In response to the community's goals, the Lifelong Mableton Project established two specific efforts designed to create opportunities for community engagement and citizen involvement in the area: an intergenerational community garden and a seasonal farmers' market. Both of these projects utilize existing community resources and create a community presence on the site of the proposed new town square identified during local planning efforts.

For example, the community garden, located in the heart of the Mableton, provides residents with a place to grow fresh fruits and vegetables, opportunities for social interaction and a location for health and wellness activities. Mableton leaders championed the concept of the garden because of its ability to foster a sense of community and to provide opportunities for community engagement. The community garden attracted significant interest from the residents, local Master Gardeners' Club, and Mableton Civic Association. The initial gardeners (40% older than 60) established a committee structure, bylaws, and a working relationship with Mableton Civic Association, a necessary process to ensure long-term sustainability of the garden. Informal surveys of the garden participants, administered throughout the first two years of operation, overwhelmingly indicated that they are choosing to garden for health and social interaction.

In addition, together with community partners, the Lifelong Mableton Project established a weekly farmers' market, which gives residents a place to purchase fresh produce and an opportunity to connect with neighbors in Mableton. The project worked with the community partners to locate the market near the garden to further support the community's vision for the town square. Community leaders felt that the market would create a sense of place and community to complement the nearby garden and help residents see activity in the same location as the future town square.

Theme 4: Addressing the Needs of Underserved Populations

The Lifelong Mableton Project worked to create a welcoming environment and ensure access for all Mableton residents, including underserved populations, throughout the planning and decision-making process to identify unmet needs. For example, the Georgia AARP State Chapter joined Lifelong Mableton as a partner in the farmers' market and community garden. Together, AARP Georgia and Lifelong Mableton provided transportation and produce coupons to resident shoppers. Also, ARC provided low-income seniors with additional assistance through vouchers through the Senior Farmers' Market Nutrition Program. Similarly, AARP Georgia targeted underserved populations as part of the community garden by incorporating "Plant a Row for the Hungry," which included sharing community garden crops of 1,000 pounds per year with a local homeless shelter. Finally, the effort provided more than 350 older adults with free vital health screenings with area health providers in nontraditional locations using the Centers for Disease Control and Prevention's Sickness Prevention Achieved Through Regional Collaboration model.

National statistics show that approximately 20% of all adults older than 55 experience mental disorders that are not part of the "normal" aging process (American Association of Geriatric Psychiatry, 2008), equating to over 1,600 older adults with mental health issues living in Mableton, including about 525 individuals who have serious mental illnesses. With the knowledge of national statistics, as well as identified local need and locally identified gaps in services, ARC created the Lifelong Mableton Mental Health Collaborative. This effort concentrated on the specific needs of older adults with mental health issues living in Mableton so that they too could effectively age in place and avoid negative outcomes such as arrests, hospitalizations, and homelessness. The Mental Health Collaborative consisted of more than 30 individuals, organizations, agencies, and businesses that address some aspect of mental health and/or older adults. Throughout the project, the collaborative worked together to identify local trends regarding mental illness among older adults, raise community awareness about mental illness, and improve access to mental health care and resources.

A PROCESS FOR REPLICATION

Over the past three years, ARC has replicated LLC approaches in eight other communities, including the city of Morrow, the neighborhood of Peachtree Hills in Atlanta, Tucker in DeKalb County, the city of Conyers, and 11 high-rises for the Atlanta Housing Authority. Four local governments have adopted resolutions supporting Lifelong Communities Principles, demonstrating strong political support. Implementation activities in other communities have led to new community gardens and farmers' markets, better sidewalks, safer intersections, zoning changes that allow for senior housing ordinances near services, transportation alternatives, and healthy living initiatives. The cities of Morrow and Kennesaw have used the Mableton walkability assessment. For its leadership in moving forward with Lifelong Communities, ARC has been recognized at the local, state, and national levels for its progressive initiative related to aging. ARC continues to provide regular technical support to local governments and civic groups aspiring to launch or expand Lifelong Communities.

As part of the Lifelong Communities process, ARC established best practices and other lessons from the Lifelong Mableton Project that other communities might utilize as they develop their own approaches:

1. There is no one prescription for a lifelong community; approaches must be designed to meet the needs of individual communities.
2. Strong leadership through both civic and government participation is essential.
3. Opportunities for community support, engagement, and feedback are encouraged at every level, from pre-planning to sustainability.
4. Early wins are necessary to help the community to visualize the Lifelong Communities Initiative.
5. A lifelong community cannot be forced upon people; rather, it must be crafted with input and guidance from those whom it is intended to serve.

Overall, the Lifelong Communities Initiative is a well-defined model and proactive approach that will support the forecasted growth of the older adult population in the region while leveraging the resources of the local communities to support broad scale implementation of the Lifelong Communities Initiative. The Lifelong Communities Initiative is a model that cuts across age groups with the notion that what works for older adults will work for the entire community. A community interested in launching a Lifelong Community effort is encouraged to learn about opportunities in its own area, convene local leaders around important issues identified by the community, and educate community stakeholders and other leaders on necessary solutions to age in place.

NOTES

1 The Carl Vinson Institute of the University of Georgia drew 11 independent samples for each of the 10 counties and the City of Atlanta within the Atlanta Regional Commission's 10-county boundary area. In total, 1,651

individuals aged 55 and older were interviewed between September 15 and September 14, 2006.

2 The first phase of the town square is currently under construction and funded through a voter approved local option sales tax.

REFERENCES

AARP. (2003). *These four walls . . . Americans 45+ talk about home and community*. Available from http://assets.aarp.org/rgcenter/il/four_walls.pdf

AARP Public Policy Institute. (2011). *Aging in place: A state survey of livability policies and practices*. Available from http://assets.aarp.org/rgcenter/ppi/liv-com/ib190.pdf

American Association of Geriatric Psychiatry. (2008). *Geriatrics and mental health—the facts. Available* from http://www.aagponline.org/prof/facts_mh.asp

Atlanta Regional Commission. (2007). *Older adults in the Atlanta region: Preferences, practices, and potential of the 55+ population*. Available from http://www.atlantaregional.com/File%20Library/Aging/ag_aging_survey_5_1_08.pdf

Ball, M. S. (2008). *Aging in place: A toolkit for local governments*. Available from http://www.atlantaregional.com/File%20Library/Aging/ag_aging_in_place_toolkit_2_23_04.pdf

Howe, D. (2001). Aging and smart growth: Building aging sensitive communities. *Funder's network for smart growth and livable communities*, 7. Available from www.fundersnetwork.org

Jacobsen, L., Kent, M., Lee, M., & Mather, M. (2011). America's aging population. *Population Reference Bureau*, 6(1), 1–18.

King, G., & Farmer, J. (2009). What older people want: Evidence from a study of remote Scottish communities. *Rural and Remote Health*, 9(2), 1166.

Knickman, J. R., & Snell, E. K. (2002). The 2030 problem: Caring for aging baby boomers. *Health Services Research*, 37, 849–884.

Masotti, P. J., Fick, R., Johnson-Masotti, A., & MacLeod, S. (2006). Healthy naturally occurring retirement communities: A low-cost approach to facilitating healthy aging. *American Journal of Public Health*, 96, 1164–1170.

MetLife Mature Market Institute. (2010). *Aging in place 2.0: Rethinking solutions to the home care challenge*. Available from http://www.metlife.com/assets/cao/mmi/publications/studies/2010/mmi-aging-place-study.pdf

National Association of Area Agencies on Aging. (2011). *Advancing health and long-term services and supports. 2010 survey of area agencies on aging*. Available from http://www.n4a.org/pdf/2011_AAA_Survey_Report.PDF

Neal, M. B., Chapman, N., Dill, J., Sharkova, I., DeLaTorre, A., Sullivan, K., & Martin, S. (2006, August). *Age related shifts in housing and transportation demand*. Available from www.pdx.edu/ims

New York Academy of Medicine. (2008). *Toward an age-friendly New York City: A findings report*. Available from http://www.health.ny.gov/diseases/aids/conferences/docs/nyam_age_friendly_report.pdf

New York Academy of Medicine. (2012). *Creating an age-friendly NYC one neighborhood at a time*. Available from http://www.nyam.org/agefriendlynyc/docs/Toolkit_Report_0321-VA-new.pdf

Philadelphia Corporation for Aging. (2004). *The socio-economic and health characteristics of Philadelphia's aging population*. Available from http://www.pcacares.org/Files/socioeconomic_and_health_keyfindings.pdf

Philadelphia Corporation for Aging. (2006). *Looking ahead: Philadelphia's aging population in 2015*. Available from http://www.pcacares.org/Files/2015_report.pdf

Philadelphia Corporation for Aging. (2008). *The Milestones survey on growing old in Philadelphia 2008*. Available from http://www.pcacares.org/Files/milestones_webSurvey.pdf

Plouffe, K. (2011). Towards global age-friendly cities: Determining urban features that promote active aging. *Journal of Urban Health*, *87*, 733–739.

Portland State University. (2009). *Creating living communities for an aging society*. Available from http://www.pdx.edu/sites/www.pdx.edu.ioa/files/Smr09_Hollywood_Cmnty_Ptnrs_final.pdf

Richard, L., Gauvin, L., Gosselin, C., & Laforest, S. (2008). Staying connected: Neighborhood correlates of social participation among older adults living in an urban environment in Montréal, Québec. *Health Promotion International*, *24*(1), 46–57.

Tinker, A. (1997). The environment of ageing. *The Royal Society*, *1997*(352), 1861–1869.

U.S. Census Bureau. (2008). *American community survey*. Available from http://www.census.gov/acs/www/data_documentation/2008_release/

World Health Organization. (2007). *Global age-friendly cities: A guide*. Available from http://www.who.int/ageing/publications/Global_age_friendly_cities_Guide_English.pdf

Ziebarth, A., Erm, H., & Hwang, E. (2010). *Designing age-friendly communities*. Available from http://www.informedesign.org/_news/march_v08.pdf

11 Making the Right Moves

Promoting Smart Growth and Active Aging in Communities

Kathleen E. Sykes and Kristen N. Robinson

> Cities have the capability of providing something for everybody, only because, and only when, they are created by everybody.
>
> (Jacobs, 1961)

> . . . to make our nation's communities good places to grow old, we must also ensure that we involve seniors themselves in the decision making and involve them in services. Older adults are a huge resource for this nation in terms of the talents, experience and capacity to give back in service.
>
> (Markwood, 2003)

INTRODUCTION

As our population rapidly ages, can communities meet the diverse requirements of older adults for housing, transportation, health, and social needs? Can local planning, policies, and programs promote health and encourage active aging? How can we incentivize communities to prepare for the aging elder population?

The goal of this chapter is to describe the winners of the Building Healthy Communities for Active Aging (BHCAA) Award Program and to review their successes and best practices. To do this, we summarized the history of the award program, examined the conceptual framework of the U.S. Environmental Protection Agency (EPA) model for smart growth and active living, and described and discussed best practices that emerged from the BHCAA award program (U.S. EPA, 2012).

Background

Between 2006 and 2011, a recognition program—BHCAA—has sought to spur towns, cities, and regional agencies to adopt and implement principles of smart growth and active aging. The BHCAA program can trace its roots to the 2002 launch of the U.S. EPA Aging Initiative. The initiative had four goals: (1) conduct research to better understand environmental health hazards that affect older people; (2) translate research findings into practical public health interventions; (3) create tools to address the impact of a rapidly growing aging population on the environment; and (4) provide opportunities for older adults to become environmental stewards in their communities.

In 2004, the U.S. EPA Aging Initiative issued a request for proposals for demonstrations on how smart growth activities could improve the quality of life for older adults while improving environmental quality (U.S. EPA, 2004). Five organizations were chosen and awarded $25,000 each to undertake these pilots. The award winners were the Atlanta Regional Commission, Berkeley-Charleston-Dorchester Council of Governments, the Denver Council of Governments, the International City/County Management Association, and Partners for Livable Communities in collaboration with the National Association of Area Agencies on Aging.

The following year, a small group of colleagues met to discuss ways to promote policy changes at the community level that would both improve the health of the environment and increase physical activity among older adults. Then a larger group of interested individuals held regular conference calls and eventually became a 14-member steering committee led by the U.S. EPA. Members of the steering committee included representatives from the Centers for Disease Control and Prevention (CDC), the President's Council for Fitness and Sports, the National Council on Aging, the National Blueprint, the Robert Wood Johnson Foundation, and the Active for Life program.

The committee's first order of business was to convene key stakeholders. In May 2005, representatives from federal agencies, national aging organizations, public health, local planning, smart growth, and city management organizations attended. The purpose of the meeting was to decide whether and how to establish an award program to highlight successful integration of smart growth and active aging. There was strong support for a recognition program managed by the U.S. EPA. Those eligible for the award would be public-sector entities, tribes, and nonprofits working with a public entity. Awards would be presented to communities that demonstrate the best and most inclusive overall approach to implementing smart growth and active aging at the neighborhood, tribal, municipality, county, and regional levels. Applicants had to demonstrate that significant activity related to a policy, project, or program had occurred within the last five years related to both smart growth and active aging.

Smart growth is defined and characterized by development patterns that create attractive, distinctive, and walkable communities that give people of varying ages, wealth, and abilities a range of safe, affordable, and convenient choices regarding where they live and how they get around. Smart growth approaches can enhance neighborhoods and involve citizens in development decisions. Growing smart also ensures that existing resources are used efficiently and that lands and buildings that shape communities are preserved. Creating a mix of housing options makes it possible for aging Americans to find safe and affordable homes. Older adults prefer to have the option to age in place.

The BHCAA award defines active aging as enabling older adults to engage in an assortment of structured and unstructured physical activities. Communities can promote active aging by creating physical activity opportunities that are accessible and self-directed. These activities should help elders increase endurance, strength, flexibility, and balance while avoiding injury. Self-directed activities include walking, biking, fitness trails, and similar activities that are appropriate for participants at various levels of fitness and functional ability.

To be considered for an award, the activity described in the application had to encourage development to address at least one of the following 10 principles of smart growth: encourage a mix of land use; take advantage of compact design; create housing options; develop walkable neighborhoods; foster distinctive, attractive neighborhoods with a sense of place; preserve open space, farmland, and critical environmental areas; strengthen and direct development to existing communities; provide transportation choices; make development decisions predictable, fair, and cost-effective; and encourage community collaboration in development decisions. Applicants must also have written a community goal to increase levels of physical activity among older adults. Applicants were also asked to discuss the following elements:

- *Effectiveness*. What are the goals of the project, policy, or program? Is there tangible evidence of "on the ground" success? How well does the activity reflect the coordination of smart growth strategies and active aging programs? Is it part of a wider strategy to encourage smart growth and active aging?
- *Outreach and community involvement*. What public outreach and education has occurred as part of the activity? How have citizens and stakeholders, especially older adults, been informed about the activity, encouraged to give input, and engaged in making decisions? How well does the activity foster collaboration among local and state governments, federal agencies, communities, nongovernmental or nonprofit organizations, or businesses and the aging network?
- *Innovation*. Does the project, policy, or program break new ground? Does it represent a new approach to a common problem? Could this new idea be adopted or replicated elsewhere? What were the key components of successful implementation?
- *Environmental and health impact*. How does the project, policy, or program improve environmental quality and help to increase active aging?

Initially, applicants needed only to include a letter of recommendation from their local area agencies on aging. Later, a letter of support was also required from their local planning organizations. We had been receiving applications that focused on either active aging or smart growth. The purpose of the recognition award was to demonstrate the benefits by combining these two concepts. By requiring letters from the local planning organizations, it was hoped they would have a better understanding of the purpose of the changing needs of community residents as they age and work with their local area agencies on aging.

Recognition Award Program: BHCAA

The BHCAA award program was launched in Washington, DC, on September 26, 2006. At the same time, a website was unveiled to inform the public and other stakeholders interested in the intersection of smart growth and active aging strategies. The purpose of the award program was to raise awareness about healthy synergies that could be achieved when communities combine smart growth and active aging strategies. Two award levels were created, the Achievement Award and the Commitment Award. The Commitment Award recognized communities that have initiated specific plans to implement smart growth and active aging principles. The Achievement Award honored communities that had successfully implemented at least one principle of smart growth and active aging. Twenty-three organizations, including the Administration on Aging and the CDC's Healthy Aging Research Network, were initial supporters and advocates of the award. By 2011, a total of 65 organizations had joined the effort to raise awareness about the recognition award.

Due to limited resources, the BHCAA award recognized winners but did not provide funds. Nor did the award program compensate the experts who annually reviewed the applications. The reviewers were experts from the environmental and aging fields who had expertise in land use, smart growth, and active aging. Due to the passion and dedication of many hard working individuals, this award program was able to thrive for five years with limited resources and only one staff person.

Between 2007 and 2011, 21 communities in 15 states received BHCAA awards (see Table 11.1). Five states boasted more than one winning community, including Georgia, Maryland, North Carolina, Texas, and Virginia. Seven communities received the Achievement Award, while 14 communities received the Commitment Award. Award applicants included local agencies, departments, and organizations; their efforts varied widely. Forty percent of the winning applicants were spearheaded by a local aging entity, such as an area agency on aging. Regional councils of governments, counties, cities, and towns were recognized for their commitments to smart growth and active aging and their accomplishments. These winning communities became ambassadors for the award program and helped other communities in their efforts. In December 2011, the U.S. EPA decided to incorporate the aging award within the equity award, one of the EPA's National Awards for Smart Growth.

Table 11.1 Building Healthy Communities for Active Aging Award (BHCAA) Winners 2007–2011

Award Recipient	State	Level	Year	Applicant Organization
City of Rogers	AR	C	2007	City Health and Wellness Center
County of San Mateo, Half Moon Bay Senior Campus Plan	CA	C	2009	Aging, Senior Housing (HUD funded)
City of Satellite Beach	FL	C	2008	Mayor
Atlanta Regional Commission (ARC)	GA	A	2007	Division of Aging Services
ARC, Lifelong Mableton	GA	C	2011	Area Agency on Aging
City of Iowa City	IA	C	2008	City Planning Department
City of Gaithersburg, Upcounty Senior Center	MD	C	2009	Senior Center
Queen Anne's County Housing Authority	MD	C	2007	County Housing Authority
Town of Scarborough	ME	C	2007	Community Services Department
Carver County. Public Health Division & Health Partnership	MN	C	2007	County Health Department and Health Partnership
City of Charlotte	NC	A	2010	Transportation
Town of Cary	NC	A	2009	City Planning Agency
Bureaus of Portland Parks & Recreation and Transportation	OR	A	2008	Bureaus of Parks/ Recreation and Transportation
Philadelphia Corporation for Aging	PA	C	2010	Area Agency on Aging
Brazos Valley Council of Governments (BVCOG)	TX	C	2007	Area Agency on Aging
Brazos Valley Council of Governments (BVCOG)	TX	A	2010	Area Agency on Aging
City of Temple	TX	C	2011	City Manager
Martinsville-Henry County Coalition for Health and Wellness	VA	A	2009	County Health Department and Health Coalition

(*Continued*)

Table 11.1 (Continued)

Award Recipient	State	Level	Year	Applicant Organization
Fairfax County. Dept. of Neighborhood & Community Services	VA	C	2010	Neighborhood and Community Services Department
City of Kirkland	**WA**	A	2007	**Parks and Community Services Department**
City of Casper	**WY**	C	2008	**City Manager**

Note. Award winners are listed in alphabetical order by state.

Level: Refers to the award levels: Achievement Award (A) Commitment Award (C).

Year: Year Awarded the BHCAA Award.

Applicant Organization: Organization(s) completing the application for the BHCAA Award.

Text in bold: Refers to the nine organizations completing the survey.

METHODS

Each community was required to submit an application that discussed the overall effectiveness of its programs, level of community involvement and outreach, use of innovative approaches, and overall environmental and health benefits of the project. In the subsequent years, previous winners were invited to provide updates that were included in the annual award booklet. Among those that submitted updates, the authors selected nine communities that best represented the winners to be interviewed for this chapter during the fall of 2012.

The survey questions were taken primarily from the initial award application. Specifically, we were interested in how the communities determined their outcome measures and whether they noticed any additional improvements after implementing their programs and policies. We asked which principles of smart growth their communities had adopted and implemented (e.g., mixed land use, walkable communities, choices for transportation, and housing). The communities responded enthusiastically and provided us with detailed information, which we present in the next section.

FINDINGS

Measuring the success of a program can be a difficult undertaking. Often programs are evaluated based on whether the interventions and services they put in place are cost-effective. With smart growth, however, the immediate benefits (i.e., keeping citizens active and healthy, improving air quality, and staying connected and engaged) are hard to measure. The longer-term benefits—extending the amount of time older people remain living in the community, preventing chronic diseases, and improving the environmental well-being and sustainability of a community—are easier to measure, but they need to be tracked over time. Therefore, to better understand the effectiveness of integrating smart growth and active aging at the community level, we asked a select group of the award winners to answer 10 questions about their programs. These questions ranged from asking about the biggest challenges the communities encountered in building a healthy community for active aging to what outcome measures were used to measure success. While this was not a cross-cutting evaluation of the programs, it did provide an update on the progress communities had made since receiving their awards. In addition, it gave us unique insight into how communities view success and what they use as their outcome measures. Not surprisingly, the questionnaire responses varied widely due to the diversity of the communities, the unique elder population of each community, and the flexibility surrounding smart growth principles and active aging programming. However, one common theme we gleaned from the respondents was that, while the award primarily focused on improving the health and well-being of older adults, it was the whole community that benefited from the design, program, and policy changes that went along with it.

Kinds of Actions

As was mentioned earlier, there are two different types of awards: the Achievement Award and the Commitment Award. Both awards require communities to demonstrate that they have implemented (or have plans to implement) the principles of smart growth together with the concept of active aging. Due to this requirement, the majority of the communities were well on their way to establishing healthy communities when they applied for the BHCAA award. Because the process of building a healthy community involves so many different components (i.e., working with multiple agencies, being responsive to numerous stakeholders, managing various funding sources), as expected, communities had multiple projects underway when they applied. Thus, as a result of receiving the BHCAA award, most communities reported that they were able to move quickly to implement new plans for continued smart growth development. In addition, several communities reported that these post-award projects were shaped by their experience of winning the award. For example, Brazos Valley Council of Governments (BVCOG), Texas, reported that, since winning the award, they have continued to incorporate street improvements, curb cuts, and community-friendly lighting into numerous projects throughout the city to make it more accessible and safe.

Principles of Smart Growth

We asked the communities what principles of smart growth their neighborhoods now embrace. Almost all the communities reported that they now promote mixed land use; take advantage of compact building designs; create walkable neighborhoods; support a range of housing opportunities and choices; foster distinctive and attractive neighborhoods that have a strong sense of place; protect open space, farmland, and critical environmental areas; direct development to existing communities; and encourage community collaboration in development decisions. The principles of smart growth that were less likely to be implemented among the surveyed communities included providing a variety of transportation choices and improving the way development decisions are made (e.g., to be more predictable, fair, and cost-effective). Only the Atlanta Regional Commission, Georgia, and Casper, Wyoming, cited these as smart growth principles that had been realized.

Outcome Measures

The outcome measures used by the communities to assess their success varied considerably. Several communities cited levels of community involvement as their primary measure of success. Others reported using participant surveys, program attendance, and number of classes taught as outcome measures. The City of Casper said that its best measure of success was seeing more organizations and agencies incorporating the needs of older adults into their strategic planning. All the communities noted the importance of measuring success not only to ascertain program effectiveness but also to determine future needs as their residents continue to age.

Most Important Change

We asked the communities what they thought was the most important change taking place in their communities. The surveyed communities listed the following:

- Town of Scarborough, Maine, approved several senior housing projects.
- County of San Mateo, Half Moon Bay Senior Campus Plan, California, representatives said that creating an environment in which affordable housing developers and service providers could work together was essential for developing a senior campus that provided a continuum of care. This campus offered care for the most frail and economically burdened, ensuring that residents could live a quality life in a caring and supportive community.
- Kirkland, Washington, representatives responded that the key factor to the well-being of older adults was their ongoing commitment to provide opportunities for active aging.
- BVCOG, Texas, representatives highlighted their ability to provide the community with knowledge, training, and confidence to self-manage

their chronic diseases; more affordable, accessible quality "senior friendly" housing developments with curb cuts and good lighting; and well-designed walking trails.

- Respondents from Casper, Wyoming, reported that they have emphasized making their community pedestrian-friendly by adding new sidewalks, trails, and pathways throughout the community.
- Respondents from Satellite Beach, Florida, said that they have recognized that proactively addressing health care issues in older residents is cost-effective. Also, calls for emergency medical services have been reduced.

The overarching theme in the communities' responses was that, due to the high costs of long-term care and the physical, mental, and emotional stress it has on older adults and their family members, keeping people residing and participating in the community for as long as possible was their number one goal. Any changes that could be made to reach this goal were worth trying.

Replication

A measure of a successful program is the number of times it is replicated. We asked the communities whether their programs had been instituted elsewhere. Two communities reported that they had been. BVCOG reported that their evidence-based programs have been replicated statewide. BVCOG's Area Agency on Aging, in partnership with Texas A&M School of Rural Public Health, participates in a Council on Government's State Planning Region that covers 45 Texas counties. This regional partnership has enabled the BVCOG to disseminate broadly evidence-based programs for elders by providing technical assistance and training of lay leaders throughout the region. Since 2010, evidence-based programs have more than doubled. The city of Kirkland, Washington, reported that other local cities had replicated many of their programs, including starting groups like the Kirkland Senior Council, implementing walking programs, and participating in regional programs to improve transportation choices for older adults. Because program replication was not part of the original criteria to receive the award, we were not surprised that only two communities reported that their programs had been replicated elsewhere.

Other Opportunities

We asked the communities whether receiving the BHCAA award has led to other opportunities to improve smart growth and active aging. This open-ended question gave the communities a chance to share some of their more exciting accomplishments. Satellite Beach's county commissioners agreed to fund a "community medic" to visit older adults on a weekly basis. The goal of having the medic was to improve the health of the community by monitoring the health status of at-risk residents and to see whether there were changes that required medical follow-up. The medic program has been well received by the community.

A local health care district in San Mateo County awarded funds to assist with the development of 20 affordable assisted living units as part of the Half Moon Bay Senior Campus Plan. Securing funding for this development was attributed to winning the award.

Kirkland attributed winning the award to several successful programs. First, Kirkland's Senior Council produced videos for older adults including "Falling into Fitness" and "Pedestrian Safety." Second, they offered a travel ambassador for older adults so they could become familiar with the community's choices of transportation. Third, they expanded their emergency preparedness programs and partnered with other community agencies, including local hospitals. Finally, they organized a housing fair that allowed the community to review several housing options available to older adults.

BVCOG attributed one of the award's requirements, the need to form "community partnerships," with helping them to work well on other projects. They said that having different partners take ownership and responsibility for implementing small pieces of a big project helped them further their mission.

Challenges

The number one challenge reported by communities was the need for more funding. All communities cited ways they could improve their programs if they had more money. Their problems ranged from not having enough money to implement specific programs to not being able to give raises to their staffs. While funding was clearly the biggest barrier, other challenges included improving education and outreach about the new healthy living opportunities; making the community accessible for both pedestrians and drivers; and finding ways to keep people active in bad weather. However, in addition to listing challenges they faced, the communities also offered solutions. For example, Kirkland solved their problem of needing to improve the way they inform and educate the community through innovative measures. They advertised their programs in local newspapers, as well as in the parks and recreation guide; used social media to post information about their programs on their city webpage; networked with local neighborhood associations; and joined several community committees.

Limitations

There are several limitations of this study worth noting. First, due to limitations in funding, a formal evaluation of the programs was not possible. Instead, we asked communities to respond to a questionnaire and provide an update on their communities' efforts to create healthy communities for active aging. Second, due to the Paperwork Reduction Act, we were limited to surveying only nine of the winners. Third, the responses were self-reported, and no systematic methodology was used to determine the accuracy of their responses. For example, reports on replication provided by the communities were taken at face value. We do not know the extent, quality, or methods of replication that were used.

Despite these limitations, it is clear that the integration of smart growth and active aging makes a measurable difference in the lives of older adults. Quantifying these differences for empirical analyses requires staff time and resources that go beyond the scope of the original award and recognition program. Ideally, future evaluations could include outcome measures such as (1) air quality; (2) Robert Wood Johnson Foundation's County Health Rankings; (3) adoption of complete street policies; and (4) length of residence in the homes of older residents. The initiatives described in the chapter are best seen as illustrations of encouraging directions that communities can take to be supportive of older residents.

Next Steps

The "next steps" the communities outlined were essentially the same for all: (1) continue the work they are doing; (2) look for ways to improve upon what they have done; and (3) find new funding sources. These communities demonstrated that when people work together toward a common goal and have the backing of their local and state government officials, change can happen. Not only did they change the physical environment within their communities, but they also changed the thought process of those making the decisions (i.e., government officials, city planners, community members). While the communities had various thoughts, ideas, and plans on how to have a positive impact on the health and well-being of older adults, they came to the same conclusion that just thinking through these issues was a positive step. And while winning the BHCAA award was viewed as a significant accomplishment for the communities, it was in no way an end point to their work.

CONCLUSIONS

The communities clearly demonstrated that strong leadership and multiple partnerships are vital for making lasting differences. They also showed that champions and leadership can come from many different places: area agencies on aging; health departments; parks and recreation centers; city government; and local planning departments. In fact, one additional measure of success could be the breadth and resolve of partners who worked together to make their communities healthier and active places to live. Without these partnerships, the programs and policies proposed by these communities could never be implemented.

The policy implications of the changes these communities achieved are promising. While it takes time to make long-lasting policy changes in the design of a community or in the adoption of an active lifestyle, other, smaller changes can be made that immediately affect older adults. For example, affordable housing options, improved sidewalks and streets, and community-wide active aging programs all contribute to improving the health and well-being of older adults from the first day they are implemented. Not only do these changes help to support older residents, but they are also beneficial to persons of all ages and abilities.

The programs and policies highlighted in this chapter suggest promising directions that communities can take to address the changing needs of their older residents. Future research is needed, however, to analyze the impact smart growth has on the lives of older adults. If data can show that older adults are able to remain in their homes longer because of changes made by their communities, then more and more communities are likely to implement similar strategies. This chapter ends with a vignette from one of the award winners and includes a brief summary at three points in time: (1) the year the community received the BHCAA award; (2) the update the community provided a year or more after receiving the award; and (3) the results from the survey. We hope this chapter will inspire communities to redesign their built environments into healthy and active places to live. Jane Jacobs was prophetic when she wrote, "Cities have the capability of providing something for everybody, only because, and only when, they are created by everybody."

REFERENCES

Atlanta Regional Commission. (2009, June). *ARC's 20-county forecast. What the future holds. Regional snapshot.* Atlanta, GA: Atlanta Regional Commission.

Jacobs, J. (1961). *The death and life of great American cities.* New York, NY: Random House.

Markwood, S. (2003, May 7). CEO of the National Association of Area Agencies on aging, at the U.S. EPA Public Listening Session, Baltimore, MD.

U.S. Census Bureau. (n.d.). *American community survey 2007–2011.* Available from http://www.census.gov/acs/www/

U.S. Environmental Protection Agency. (2004). *Protecting the health of older adults by improving the environment: Training, innovation, outreach and educational projects.* Available from http://www.epa.gov/fedrgstr/EPA-GENERAL/2004/April/Day-16/g8678.htm

U.S. Environmental Protection Agency. (2012). *Building healthy communities for active aging national recognition program.* Available from http://www.epa.gov/aging/bhc/about.htm

Appendix

Vignette of Atlanta Regional Commission BHCAA Award Winning Community

We have included a vignette of the Atlanta Regional Commission (ARC). We include three points in time: 2007, when they received their first award; an update in 2008; and then survey results from 2012. We included information on size, population, counties for the regional winners, median household income, and poverty figures, including the overall poverty rate, as well as poverty rates in the young and the old. Data were taken from the U.S. Census American Community Survey 5-Year Estimates (U.S. Census Bureau, n.d.).

ARC COUNTIES

Cherokee, Clayton, Cobb, DeKalb, Douglas, Fayette, Fulton, Gwinnett, Henry, Rockdale, and City of Atlanta. Area: 3,024 sq. miles. Population: 4,107,750. Range of median income in ARC counties: Clayton County: $37,050 to Fayette County: $72,962. Poverty: 13.2% all people: 9.3% 65 and older; 18.9% younger than 18. Within ARC's counties, the 65 and older population ranged from 6.9% in Gwinnett County to 13.9% in Fayette County (the state average was 10.7 %).

ACHIEVEMENT AWARD 2007 (ARC), COMMITMENT AWARD 2011 LIFELONG MABLETON

In 2002, ARC assembled the necessary partners to create "Aging Atlanta." The mission of this 50-organization partnership was to meet the needs of the region's growing aging population. Between 2005 and 2040, ARC's 65+ population was predicted to grow by 318% and the 85+ population by almost 500% (Atlanta Regional Commission, 2009.) Aging Atlanta pilot projects laid the foundation for the Lifelong Communities Initiative. The initiative works with local governments to create housing and transportation options that enable older adults to age in place. ARC facilitated zoning policy changes and the development of 30 senior housing developments located close to services and connected to existing neighborhoods. With more than 90% of Atlanta's elders relying on cars for transportation, ARC took steps to decrease auto dependency by promoting ride sharing through its six voucher programs and worked to improve bus stops and routes. These efforts increased the quality of life for elders and benefitted the environment. To encourage healthy lifestyles, ARC partners converted traditional senior centers to wellness centers, emphasizing physical activity and social interaction. In 2007, 46 senior centers offered programs for the 400,000 older adults in the metro area. Approximately 1,000 individuals have joined walking clubs. Through community involvement, ARC listened to the needs of older adults and integrated age-appropriate features into parks, trails, and pedestrian paths. They also conducted sidewalk audits to ensure that local plans incorporated changes that would better serve older adults.

UPDATE 2008

ARC's Lifelong Communities Initiative has made significant progress toward meeting its goals to prepare for the growing older adult population by developing priorities, policies, and programs that support three goals: promote housing and transportation options, encourage healthy lifestyles, and expanding information and access to services. ARC and its partners made significant progress since 2007 toward all three goals, including passage of several zoning ordinances that encourage a diverse housing stock; integration of transportation needs into local comprehensive plans; and a program to train older adults how to advocate for changes to zoning codes and regulations regarding the built environment. Six of the county and city teams have formed a preventative health collaborative. In one community, ARC is working to develop a more pedestrian- and elder-friendly downtown. The initiative has been able to bolster existing and initiate new walking clubs, helping more elders increase their physical activities.

SURVEY RESULTS 2012

Since winning the Achievement Award in 2007 for their Lifelong Communities Initiative and a Commitment Award in 2011 for Lifelong Mableton, ARC has been busy replicating policies and best practices in eight other communities including the city of Morrow, the neighborhoods of Peachtree Hills in Atlanta, Tucker in DeKalb County, the city of Conyers, and at 11 high-rises for the Atlanta Housing Authority. Three local governments, DeKalb, Clayton, and Avondale Estates, have adopted resolutions supporting Lifelong Communities, demonstrating strong political support. The establishment of new multidisciplinary stakeholder committees resulted in greater collaboration among local officials and residents.

ARC believes that the most important change that Lifelong Mableton Project has made is the understanding that to create a truly sustainable Lifelong Community, the initiative must be owned and led by the community, which fosters leadership and involvement. Maximizing partnerships continues to be the hallmark of this effort, and project leaders continue to reach into their communities and challenge organizations to see themselves as integral parts in supporting their aging communities. The Lifelong Mableton Project has found that existing structures and existing foundations are able to advance the goals of lifelong communities.

In the summer of 2010, more than 150 residents of Mableton gathered for 7 days to develop a unified vision for their community. In late February 2011, their elected officials joined together and unanimously voted to support the "Create Mableton Plan." The plan will help redefine the core of Mableton, a suburban area of 23,000 residents, into a walkable and livable community. It uses form-based code as a tool to help property owners in the community independently develop their properties based on design guidelines rather than restrictive land uses. The design regulations incorporated in the code will foster a sense of community for future generations and establish zoning regulations that focus on the design of the community to support walkable and livable spaces, promote housing choices, and improve both public and private spaces. The community is now realizing the implementation of its vision through the construction of a new state-of-the-art elementary school, located in a core redevelopment area, which anchors the community and provides opportunities for new families to move into the area. As the Lifelong Mableton Initiative progresses into its final phases, a key focus of all work efforts will include steps to ensure long-term sustainability. For the purposes of this Initiative, ARC defines the sustainability of Lifelong Mableton as the existence of structures and processes that allow a program to leverage resources to most effectively implement evidence-based policies and activities over time.

12 Does the Village Model Help to Foster Age-Friendly Communities?

Andrew E. Scharlach, Joan K. Davitt, Amanda J. Lehning, Emily A. Greenfield, and Carrie L. Graham

INTRODUCTION

Age-friendly community models primarily focus on government-initiated planning efforts designed to modify cities' physical and social infrastructures. Considerably less attention has been given to the potential role of nongovernmental organizations in promoting community-wide age-friendly efforts. One social organization that has received a great deal of attention since it was first developed in 2001 is the Village model, a membership-based grassroots initiative that facilitates social engagement, as well as access to services (Scharlach, Graham, & Lehning, 2012). Using data from a national survey, this chapter explores the potential role of Villages in fostering age friendliness within organized groups, as well as within the overall community.

BACKGROUND

The past decade has seen a dramatic increase in efforts to assist individuals to age in place. This focus on aging in place reflects older adults' expressed desire to continue to live in familiar homes and neighborhoods while meeting physical, psychological, and social needs to the extent possible (Kochera & Guterbock, 2005) and societal goals of reducing unnecessary health care and long-term care expenditures by "rebalancing" financing and service-delivery systems to promote community-based care (Doty, 2010; O'Shaughnessy, 2013). Formal efforts to promote aging in place primarily involve traditional aging network services (e.g., nutrition, transportation, case management, in-home care) provided directly to individual older persons by formal service providers, including local government, not-for-profit organizations, or private for-profit companies. The existing system of aging services has been criticized as being fragmented, uncoordinated, duplicative, and isolated from existing community social resources, contributing to inefficiency and substantial unmet needs among disabled older adults (Harrington, Ng, Kaye, & Newcomer, 2009; Institute of Medicine, 2001). Furthermore, the effectiveness of existing services is constrained by environmental barriers and infrastructure limitations, which can exacerbate functional limitations (Pynoos, Nishita, Cicero, & Caraviello, 2008; Woolf & Braveman, 2011).

At the community level, prominent efforts to promote aging in place have involved age-friendly city initiatives—community-wide projects focused on changing physical and social environments to make cities more livable and removing barriers to social participation for older adults (World Health Organization [WHO], 2007). These initiatives are typically spearheaded by local governmental agencies or philanthropic organizations, target communities as a whole, involve community-wide needs assessment and planning processes, and are designed to collectively benefit older adults and other residents by modifying community infrastructures and basic services (Lehning, Scharlach, & Price-Wolf, 2012). Prominent examples of age-friendly city initiatives include the WHO Global Network of Age-Friendly Cities and Communities program, the Visiting Nurse Society of New York's AdvantAge Initiative, the U.S. Environmental Protection Agency's Building Healthy Communities for Active Aging initiative, and AARP's Livable Communities initiative. Community-wide planning efforts such as these face a number of challenges, including inconsistent plan implementation, lack of community structures to facilitate translation from planning reports to ground-level actions, a focus on major system changes that typically occur very slowly, and inadequate inclusion of older consumers in planning and implementation processes. For example, while age-friendly city initiatives may include older adults among their key advisors, they rarely have older adults actively engaged in planning or implementation (Lehning et al., 2012).

Age-friendly community efforts are only beginning to recognize the potential benefits of actively engaging community social organizations, including membership associations, neighborhood groups, and organizational collaborations. Yet such efforts comprise the majority of community aging initiatives identified in a recent national study (Lehning et al., 2012). By engaging existing community social organizations, age-friendly initiatives can draw upon available social capital within existing communities of common interest (Halpern, 2005), facilitate older adults' engagement in the change process, and ensure that interventions are targeted most effectively to the needs and preferences of specific community sectors. Community social organizations are of increasing importance in the context of increasing population diversity and declining roles of families, religious institutions, and governments in the postmodern state (Feinberg, Reinhard, Houser, & Choula, 2011; Salamon, 2012; Szinovacz & Davey, 2008).

One of the most prominent emerging social initiatives is the Village model. Villages are membership associations developed and operated by older community members for the primary purpose of enhancing their quality of life and ability to age in place. In exchange for membership dues, participants gain access to an array of social, educational, and recreational activities; assistance with driving, housekeeping, and other support services; a dedicated source of information and assistance; and referrals to community service providers, often at a reduced rate (Scharlach et al., 2012). A unique characteristic of the Village model is the extent to which it promotes consumer engagement and social support, typically by involving older adults in governance roles and as direct providers of supportive services (McDonough & Davitt, 2011). In so doing, Villages provide an organizational framework through which older adults might together enhance their own well-being and facilitate access to existing services, while also fostering infrastructure changes that enhance age friendliness for members as well as other community residents (Greenfield, Scharlach, Lehning, & Davitt, 2012). By involving members in these activities, Villages may serve as a potential mechanism for promoting the empowerment of older adults. As noted by Irene Hoskins, past president of the International Federation on Aging, "Villages may be ideal conduits to lead efforts towards an age-friendly city" (Hoskins, 2012).

A growing consensus is emerging regarding the components of an age-friendly community (Lui, Everingham, Warburton, Cuthill, & Bartlett, 2009). The WHO Global Network of Age-Friendly Cities and Communities program, for example, identified eight domains contributing to health, participation, and security as people age, based on prior conceptual and empirical findings of the AdvantAge Initiative and the AARP Livable Communities initiative (WHO, 2007). Social domains include social participation (engagement in recreation, socialization, cultural, educational, and spiritual activities); civic participation and employment (opportunities for civic engagement, unpaid work, and paid work); and respect and social inclusion (attitudes of the community as a whole toward older people). Supportive service domains include community support and health services (access to social services as well as a range of health services that are not strictly medical) and communication and information (access to information and technologies that enable elders to stay connected and obtain needed information). Physical infrastructure domains include transportation (ability to get to places when needed); housing (ability to age comfortably and safely within one's chosen community); and outdoor spaces and buildings (environments that promote safety and accessibility).

In this study, we use data from a 2012 national organizational survey to examine the extent to which Villages help their members access supports and services consistent with the eight domains identified by the WHO Global Network of Age-Friendly Cities and Communities program, while potentially also enhancing the age-friendliness of the entire community.

METHODS

Sample and Procedures

This study used data from Villages participating in a larger national study of community-based initiatives (see www.agingandcommunity.org for more information). The sampling frame for the study's analytic sample was developed from organizations listed on the website of the Village to Village Network (http://www.vtvnetwork.org/), of which 80 were deemed eligible for participation (i.e., organizations that currently provided services to older adults and were at least partially consumer-driven). Of the 80 Villages, 69 participated, yielding a response rate of 86.3%. Of these 69 Villages, 29% served a single neighborhood, 19% a single town or city, and 42% multiple towns or cities. The Village's catchment area was reported to be of low or low to middle socioeconomic status (SES) by 20% of respondents, middle SES by 23%, and middle to high or high SES by 57%. Approximately 10% of Village members were reported to be younger than 65; 32% were aged 65 to 74; 40% were aged 75 to 84; and 18% were aged 85 or older. Twenty-four percent of members reportedly needed assistance with household chores, and 14% with personal care. Twelve percent were considered to be impoverished and another 12% economically insecure.

A representative from each Village, typically the executive director or board president, was contacted between January and June 2012 and asked to participate in the study. Participants received by e-mail a 24-page questionnaire, which included sections assessing organizational development, governance structures, financing, collaborations, programs and services, membership characteristics, member involvement, and sustainability. Each participant then engaged in an hour-long telephone interview with a trained student interviewer at the University of California at Berkeley to review and clarify their responses and answer additional questions. With the support of a grant to Rutgers University from the Silberman Fund for Social Work Faculty Research, participants received $60 as a token of appreciation for completing the interview. Because the study focused on organizations, it was deemed exempt by the Rutgers University Institutional Review Board.

Measures

Measures were selected to operationalize the eight types of age-friendly community supports identified by the WHO Global Network of Age-Friendly Cities and Communities program: social participation; civic participation and employment; respect and social inclusion; community support and health services; communication and information; transportation; housing; and outdoor spaces and buildings. Questions assessed whether relevant services and supports were provided by the Village, either directly (by Village members or staff) or indirectly (through referrals to external agencies or organizations). Responses were tabulated, and descriptive statistics were computed.

Social Participation

Engagement of participants in recreation, socialization, and related activities was assessed with three questions: (1) whether strengthening older adults' social relationships and reducing social isolation is the most or second most important goal of the organization relative to three other potential goals (enhancing service access, helping the general community to become more aging-friendly, and promoting older adults' contributions to their community; yes/no); (2) whether social, cultural, or recreational gatherings (e.g., art classes or outings) were provided to members in the past year (yes/no); and (3) the percentage of members who participate in social, cultural, recreational, or other group events in a typical month.

Civic Participation and Employment

Opportunities for citizenship, unpaid work, and paid work were assessed with five questions: (1) whether promoting older adults' contributions to their community (e.g., volunteering) is the most or second most important goal of the organization relative to other goals (yes/no); (2) whether volunteer opportunities were offered to members in the past year (yes/no); (3) the importance given to engaging members as volunteers for achieving Village goals (1 = "not at all," 5 = "extremely"); (4) the percentage of members who volunteered in a typical month; and (5) whether employment placement assistance was provided to members in the past year (yes/no).

Respect and Social Inclusion

Efforts to change attitudes toward older community members were assessed with two questions: (1) whether helping the general community to become more aging-friendly (e.g., changing the physical environment or social policies or programs) is the most or second most important goal of the organization relative to other goals (yes/no) and (2) whether the organization had advocated for changes in the community as a whole in the past year (yes/no). Specific activities identified by respondents were reviewed by two investigators, and two activity types with potential for altering community attitudes toward older people were identified: (a) public presentations or other educational events designed to increase community awareness of issues affecting older persons and (b) lobbying elected officials or participating in policymaking bodies designed to promote the well-being of older adults. Responses were coded (yes/no) in terms of their applicability to each of these two categories.

Community Support Services

Access to a range of social services was assessed with five questions asking whether each of the following services was provided to members in the past year (yes/no): (1) friendly visiting; (2) reassurance phone calls; (3) financial or money-management services; (4) legal services; and (5) professional care coordination or care-management services. Participants also were asked whether the organization had developed a list of preferred providers (yes/no) and, if so, (a) the number of providers on the list and (b) the percentage of providers offering services at a discount.

Health Services

Access to a range of medical and nonmedical health services was assessed via seven questions asking whether the following services were provided to members in the past year (yes/no): (1) health or medical services (e.g., nutritionist, occupational therapist); (2) mental health counseling; (3) health care advocacy; (4) chronic disease self-management or other health education classes; (5) preventive health screenings (e.g., blood pressure checks); (6) physical activity or exercise groups; and (7) home health care or personal care.

Communication and Information

Assistance staying connected and getting timely information to meet personal needs was assessed with three questions: (1) whether the organization had provided a central phone number or e-mail address to contact for information or assistance in the past year (yes/no); (2) whether on-call assistance was available 24/7 in the past year (yes/no); and (3) whether assistance with in-home technologies (e.g., emergency response systems) was provided to members in the past year (yes/no).

Transportation

Assistance getting to places out of walking distance was assessed by asking whether transportation assistance was provided to members in the past year (yes/no) and who provided transportation services (i.e., members, staff, or external providers).

Housing

Supports to enable living environments to be safer and more comfortable were assessed with two questions regarding provision to members in the past year (yes/no) and who provided these services: (1) home repair or maintenance services and (2) cleaning or housekeeping.

Outdoor Spaces and Buildings

Organizational activities with the potential to make the built environment more aging-friendly were assessed by asking whether the organization had advocated for physical or social changes in the community as a whole in the past year (yes/no), as described above. Specific responses were reviewed by two investigators, who identified categories reflecting actions designed to promote creation of environmental changes or infrastructure improvements of potential benefit to Village members as well as other community residents (e.g., sidewalk benches, crosswalks) and then coded each response accordingly.

RESULTS

Findings address the extent to which Villages provided access to each of the age-friendly community support types identified by WHO's Global Network of Age-Friendly Cities and Communities program.

Age-Friendly Community Supports

Social Participation

Strengthening older adults' social relationships and reducing social isolation was a primary (25%) or secondary (54%) goal of 79% of responding Villages. Social, cultural, or recreational gatherings were offered by 93% of Villages, with an average of 34% of Village members participating on at least a monthly basis.

Civic Participation and Employment

Promoting older adults' contributions to their communities was a primary (3%) or secondary (14%) goal of 17% of Villages. Engaging members as volunteers was considered extremely important to organizational success by 51% of Villages and very important by 26%. Volunteer opportunities were offered by 86% of Villages, with an average of 24% of members volunteering in a typical month. Employment placement assistance was offered by only 4% of Villages.

Respect and Social Inclusion

Thirty-eight percent of Villages advocated for changes within the community in the past year, although helping communities to become more age-friendly was a primary goal for only 2% of Villages and a secondary goal for 9%. Six percent of Villages sponsored public presentations or other educational events designed to enhance public awareness of issues affecting older persons, including accessibility awareness presentations, community education about elder abuse, and a statewide senior awareness campaign. Twenty-two percent of Villages advocated with elected officials or engaged in other political actions designed to promote the creation of policies responding to the needs of older adults in the community. Policies included enhancements to social services, housing design options, services for deaf persons, tax relief for low-income elder homeowners, property tax rebates, and increased funding for municipal aging services; policymaking bodies included municipal commissions on senior issues, transportation committees, and a chamber of commerce seniors committee.

Community Support Services

Friendly visiting was offered by 87% of Villages; reassurance phone calls were offered by 83%; financial or money management services by 58%; legal services by 55%; and professional care coordination or care management by 77%. Ninety-one percent of Villages had a list of preferred external service providers, with an average of 72 providers on their lists, and 42% offered services at a discount. Outside services most often provided at a discount included home maintenance/repair (67%), home health/personal care (59%), technology assistance (45%), and housecleaning (45%).

Health Services

Health or medical services (e.g., nutritionist, occupational therapist) were offered by 48% of Villages; mental health counseling was offered by 36%; advocacy with health care providers by 58%; chronic disease self-management or other health education classes by 54%; preventive health screenings (e.g., blood pressure checks) by 39%; physical activity or exercise groups by 73%; and home health or personal care by 80%.

Communication and Information

A central phone number or e-mail address to contact for information or assistance was offered by 99% of Villages; on-call assistance available any time, day or night, was offered by 29%; and assistance with informational or communication technologies (e.g., computers and cell phones) was offered by 91%.

Transportation

Transportation assistance was offered by 99% of Villages. Of the Villages that provided transportation support, 85% indicated that they used member volunteers to provide transportation assistance; 62% also used non-member volunteers; and 47% indicated that transportation was provided by Village staff. Some Villages also provided transportation assistance by referral to independent vendors, with discounts (31%) and without discounts (34%).

Housing

Home repair and maintenance services were offered by 97% of Villages, and house cleaning and housekeeping by 81%. Villages used staff (10%), member volunteers (60%), and other community volunteers (43%) to provide home maintenance/repair services directly. They also made referrals to independent providers with discounts (69%) and without discounts (61%). Villages reported making referrals to independent vendors for cleaning services with discounts (55%) and without discounts (46%), while fewer than 25% reported using member or community volunteers to provide cleaning.

Outdoor Spaces and Buildings

Ten percent of Villages advocated for specific improvements in the physical environment of potential benefit to Village members and other community residents. Among the activities mentioned were the following: getting a citizens' association to install sidewalk benches, making crosswalks more pedestrian-friendly, developing alternative housing options within the community, and advocating for creation of a community senior center.

DISCUSSION

This study provides initial evidence regarding ways in which Villages implement features of age-friendly community models with select groups of older community members, as well as the extent to which they may contribute to the age friendliness of the larger community. Most of the Villages examined here provided supports and services consistent with the eight domains identified by the WHO Global Network of Age-Friendly Cities and Communities program (WHO, 2007). Unlike traditional aging services providers, who typically provide only one or two types of services, 85.5% of Villages provided assistance with at least six of the WHO domains. The WHO framework is particularly relevant because it is based on prior empirical evidence, has been used to guide focus group discussions with community residents and other stakeholders, and is being implemented in numerous communities throughout the world.

Villages and other organized social groups are apt to have a number of potential benefits as providers of aging services, including reliance on a combination of paid and unpaid care providers. In the Villages studied here, for example, non-professional services (e.g., driving, grocery shopping) were more likely to be provided by social network members than by paid staff or formal service providers. Such efforts draw upon the social capital developed within existing social networks, including bonding capital (e.g., peer support capacity), bridging capital (e.g., utilization of community volunteers), and linking capital (e.g., increased member knowledge of available services and benefits). Social organizations may be especially important partners in a time of budgetary constraints, decreasing service delivery and funding roles for central governments, and increasing population diversity (Musick & Wilson, 2008). Furthermore, social engagement and social inclusion are integral to providing supports within organized social networks, reflecting the "aging in community" model proposed by Thomas and Blanchard (2009), which conceives older adults as proactive agents who share responsibility for helping to create mutually beneficial communities. The ability to target interventions to the characteristics of a specific community sector is apt to increase the probability of success, especially when members are engaged in the process of identifying and responding to the expressed needs and service preferences of their peers. Furthermore, group modalities have been shown to be especially effective for a variety of services, including health promotion, disease management, and other health-related interventions (Lorig et al., 1999).

Villages exist to serve the utilitarian group interests of their members, as do most other membership organizations, professional associations, labor unions, religious organizations, and political parties. However, such organizations also have the potential to produce public goods that are beneficial to all community members (Welzel, Inglehart, & Deutsch, 2005). Social networks can serve as a collective resource that enables individuals to engage in joint actions for common goals more effectively than acting alone (Coleman, 1988; McAdam, Tarrow, & Tilly, 2001; Welzel et al., 2005). Advocacy efforts flow from members' social relationships, reflect the interests and needs of the members, and have the potential to enhance individual and group self-efficacy and social capital. In facilitating coordinated collaborative efforts to improve environmental conditions, Villages and other organized initiatives serving targeted subgroups of community members can also foster members' feelings of agency and connectedness, thereby potentially contributing to enhanced physical and psychological well-being (e.g., Wahl, Iwarsson, & Oswald, 2012; Wiles, Leibing, Guberman, Reeve, & Allen, 2012).

While community age friendliness was not a primary organizational goal for most Villages, more than one-third of the Villages in this study were engaged in some kinds of efforts to make their communities more age-friendly, including specific enhancements to the physical and social environment of potential benefit to Village members and other community residents. Environmental improvements such as these can help to reduce the need for assistance and thereby enhance the ability of individual elders to age in place, while also having direct and indirect benefits for current and future elders and the community as a whole. However, only one-tenth of Villages were working to foster specific physical environment improvements, and many of these change efforts were fairly limited, such as to improve the safety of one intersection. Community-wide initiatives targeting multiple aspects of the physical environment are likely to require the involvement of local governments, which are better positioned to enhance the age-friendliness of the physical environment through zoning changes, housing incentives, transportation improvements, and other infrastructure enhancements. Villages may have untapped potential to serve as partners with other organizations in engaging older adults in efforts to promote community-wide age-friendliness.

About one-fourth of Villages were working to improve community attitudes toward older persons, either through public presentations or by promoting development of senior-friendly public policies. Such efforts may increase community awareness of the needs and abilities of older residents, thereby helping to "make aging issues a community-wide priority" (Feldman & Oberlink, 2003, p. 269). In addition, Villages may have the potential to enable communities to benefit more fully from the wisdom and untapped energies of their older residents by facilitating opportunities for members to participate in volunteer activities, paid work, and other types of civic engagement; however, our findings provide relatively limited evidence that Villages actually are assuming that role so far.

The quality and affordability of community aging services systems also may be enhanced through the shared purchasing power of organized groups of older persons. Nearly 40% of Villages had negotiated with external service providers to serve their members at a discount, and 71% monitored the services and supports that members received. Although the current study did not examine actual impacts on service cost or quality, it seems likely that Village members' consolidated purchasing power might enable them to negotiate for better quality services at a lower cost, with potential secondary impacts on the quality of goods and services available to other older adults living in the area. Consolidated purchasing power combined with feedback mechanisms to monitor the quality of services received might help to incentivize providers to improve service effectiveness and responsiveness, potentially resulting in improvements in service system quality, accessibility, efficiency, affordability, and effectiveness.

Villages and other community organizations target limited segments of the population. Villages in this study, for example, included members who were younger, less economically insecure, and less likely to reside in communities with low or middle socioeconomic status than the general U.S. population aged 65 and older (Administration on Aging, 2012; Greenfield, Scharlach, Lehning, Davitt, & Graham, 2013). However, despite their rather limited focus, additional analyses revealed that 87% of Villages indicated that they had provided services to non-members in the past year, representing an average of 15% of all program participants. Even though some Villages may offer assistance to non-members as a recruitment strategy, providing services to non-members may enable Villages to reach parts of the larger community, while potentially reducing the demand on existing community programs and services.

The actual benefits of a social initiative may vary with the type of social organization and the group it serves. Our focus here has been on Villages. However, it is possible that other organized community groups (e.g., churches, affinity groups, other voluntary sector networks) may also have substantial potential for enhancing their members' ability to age in place while also promoting constructive changes in the overall community. A focus on existing community groups may help to expand access to community-level options to age in place in under-resourced areas.

Policy and Programmatic Implications

The United States already has among the highest rates of voluntary association membership in the world (Schofer & Fourcade-Gourinchas, 2001). New forms of intentional social organizations are emerging, in part in response to declines in the stability of traditional family and informal social networks and in their ability to meet the needs of increasing numbers of older community members from diverse communities (Feinberg et al., 2011). These emerging organizations have substantial potential for addressing the well-being of others, rather than only serving members' personal or parochial interests (Welzel et al., 2005).

Our findings suggest that community social organizations may have an important role to play in efforts to foster age-friendly communities (Scharlach & Lehning, 2013). A more integrated approach, drawing upon existing social capital to translate community ties into collective action, might enhance the ability to meet the existing needs of targeted groups of older adults, while fostering prevention-oriented efforts to make communities more age-friendly in general (Coleman, 1988; Welzel et al., 2005). Social policies and community planning efforts need to consider incentives and supports for engaging existing social organizations and membership associations in community-wide age-friendly initiatives. New York City's Aging Improvement Districts, part of the city's Age-Friendly NYC initiative, for example, are attempting to implement a citywide plan at the local neighborhood level, bringing together a wide range of community stakeholders to facilitate cross-sectional coordination and collaboration (New York Academy of Medicine, 2012).

Limitations and Directions for Future Research

While the study sample included 86% of operational Villages known to the Village-to-Village Network, no comparable information is available regarding the non-respondents, so we are unable to generalize our findings to all Villages. More important, Villages are a relatively new organizational model that is rather unique, to the extent that they represent intentional social networks with substantial consumer engagement in organizational development, governance, and service provision. The role of more traditional forms of social organizations (e.g., churches, fraternal associations) in supporting aging in place and age-friendliness across more diverse geographic communities is an important area for future research.

While the WHO Global Network of Age-Friendly Cities and Communities framework was selected as a benchmark for the purposes of this study, other models of age-friendliness may have yielded somewhat different findings. Operationalization of the variables did not permit a fully inclusive examination of all of the features identified in the WHO framework; for example, the survey did not include established housing indicators, such as the availability of a range of housing options or assistance in accessing available options. More in-depth exploration of the full array of age-friendly characteristics is needed, with particular attention to the relationship between community infrastructure change and individual aging in place services and supports.

The findings reported here identify Villages' activities designed to serve their members and sometimes affect their communities, but leave unanswered the question of whether these efforts actually contributed to members' health, participation, security, ability to age in place, or other meaningful individual outcomes or to actual environmental changes or enhanced community age friendliness. As others have noted previously, "there has been limited documentation of the effectiveness of specific approaches [for promoting age-friendly communities] or evaluation of the impact of specific processes or outcomes on older people's lives" (Lui et al., 2009, p. 119). Existing social organizations and emerging program models serving targeted community subgroups (e.g., Villages) may have an important role to play in collaboration with community-wide aging-friendly initiatives and traditional aging network services. Needed now is additional research that examines the added value of social network interventions such as Villages for promoting the well-being of network members, as well as community residents of all ages, and their potential for system-wide cost savings.

REFERENCES

Administration on Aging. (2012). *A profile of older Americans: 2011*. Washington, DC: U.S. Department of Health and Human Services. Available from http://www.aoa.gov/AoARoot/Aging_Statistics/Profile/index.aspx

Coleman, J.S. (1988). Social capital in the creation of human capital. *American Journal of Sociology, 94*, 95–121.

Doty, P. (2010). The evolving balance of formal and informal, institutional and non-institutional long-term care for older Americans: A thirty-year perspective. *Public Policy & Aging Report*, *20*(1), 3–9.

Feinberg, L., Reinhard, S., Houser, A., & Choula, R. (2011). *Valuing the invaluable: 2011 update, the growing contributions and costs of family caregiving*. Washington, DC: AARP Public Policy Institute.

Feldman, P., & Oberlink, M. (2003). The AdvantAge initiative: Developing community indicators to promote the health and well-being of older people. *Family & Community Health*, *26*, 268–274.

Greenfield, E. A., Scharlach, A., Lehning, A., & Davitt, J. (2012). A conceptual framework for examining the promise of the NORC Program and Village models to promote aging in place. *Journal of Aging Studies, 26*, 273–284.

Greenfield, E.A., Scharlach, A.E., Lehning, A.J., Davitt, J.K., & Graham, C.L. (2013). A tale of two community initiatives for promoting aging in place:

Similarities and differences in the national implementation of NORC programs and Villages. *The Gerontologist, 53*, 928–938.

Halpern, D. (2005). *Social capital*. Cambridge, UK: Polity Press.

Harrington, C., Ng, T., Kaye, S., & Newcomer, R. (2009). *Home and community-based services: Public policies to improve access, costs and quality*. San Francisco, CA: UCSF Center for Personal Assistance Services.

Hoskins, I. (2012, October 17). *Aging locally and aging globally: Where do we position the innovation of Villages?* Presentation at the 4th Annual National Village Gathering, Atlanta, GA.

Institute of Medicine. (2001). *Improving the quality of long-term care*. Washington, DC: National Academies Press. Available from http://iom.edu/Reports/2000/Improving-the-Quality-of-Long-Term-Care.aspx

Kochera, A., & Guterbock, T. (2005). *Beyond 50.05. A report to the nation on livable communities: Creating environments for successful aging*. Washington, DC: AARP.

Lehning, A., Scharlach, A., & Price Wolf, J. (2012). An emerging typology of community aging initiatives. *Journal of Community Practice, 20*, 293–316.

Lorig, K. R., Sobel, D. S, Stewart, A. L., Brown Jr., B. W., Ritter, P. L., González, V. M., . . . Holman, H. R. (1999). Evidence suggesting that a chronic disease self-management program can improve health status while reducing utilization and costs: A randomized trial. *Medical Care, 37*, 5–14.

Lui, C., Everingham, J., Warburton, J., Cuthill, M., & Bartlett, H. (2009). What makes a community age-friendly: A review of international literature. *Australasian Journal of Ageing, 28*, 116–121.

McAdam, D., Tarrow S., & Tilly C. (2001). *Dynamics of contention*. Cambridge, MA: Cambridge University Press.

McDonough, K. E., & Davitt, J. K. (2011). It takes a village: Community practice, social work, and aging-in-place. *Journal of Gerontological Social Work, 54*, 528–541.

Musick, M., & Wilson, J. (2008). *Volunteers: A social profile*. Bloomington, IN: Indiana University Press.

New York Academy of Medicine. (2012). *Creating an age-friendly NYC one neighborhood at a time*. Available from http://www.nyam.org/agefriendlynyc/docs/Toolkit_Report_0321-VA-new.pdf

O'Shaughnessy, C. (2013). *The basics: National spending for long term services and supports (2011)*. Washington, DC: National Health Policy Forum, George Washington University.

Pynoos, J., Nishita, C., Cicero, C., & Caraviello, R. (2008). Aging in place, housing, and the law. *Elder Law Journal, 16*, 77.

Salamon, L. (2012). The resilient sector: The future of nonprofit America. In L. Salamon (Ed.), *The state of nonprofit America* (2nd ed., pp. 13–19). Washington, DC: Brookings Institution Press.

Scharlach, A., Graham, C., & Lehning, A. (2012). The "Village" model: A consumer-driven approach for aging in place. *The Gerontologist, 52*, 418–427.

Scharlach, A., & Lehning, A. (2013). Aging-friendly communities and social inclusion in the United States. *Ageing and Society*, *33*, 110–136.

Schofer, E., & Fourcade-Gourinchas, M. (2001). The structural contexts of civic engagement: Voluntary association membership in comparative perspective. *American Sociological Review, 66*, 806–828.

Szinovacz, M. E., & Davey, A. (2008). The division of parent care between spouses. *Ageing and Society, 28*, 571–597.

Thomas, W., & Blanchard, J. (2009). Moving beyond place: Aging in community. *Generations, 33*(2), 12–17.

Wahl, H.-W., Iwarsson, S., & Oswald, F. (2012). Aging well and the environment: Toward an integrative model and research agenda for the future. *The Gerontologist, 52*, 306–316.
Welzel, C., Inglehart, R., & Deutsch, F. (2005). Social capital, voluntary associations and collective action: Which aspects of social capital have the greatest 'civic' payoff? *Civil Society, 1*(2), 121–146.
Wiles, J., Leibing, A., Guberman, N., Reeve, J., & Allen, R. (2012). The meaning of "aging in place" to older people. *The Gerontologist, 52*, 357–366.
Woolf, S., & Braveman, P. (2011). Where health disparities begin: The role of social and economic determinants and why current policies may make matters worse. *Health Affairs, 30*, 1852–1859.
World Health Organization (WHO). (2007). *Global age-friendly cities: A guide*. Geneva, Switzerland: WHO Press.

13 Multigenerational Planning

Integrating the Needs of Elders and Children

Mildred E. Warner and George C. Homsy

INTRODUCTION

Literature on local government response to aging, including some of the other chapters in this volume, is replete with excellent examples of the ways that municipalities reform their communities—either physically or programmatically—to make them more welcoming of older adults. The aging of the American population requires local leaders to create more supportive environments for older citizens. While this represents an important movement in urban planning and service provision, it reinforces the traditional mode of government policymaking, which treats each segment of a community within a separate silo, in this case age-segregated silos.

In this chapter, we explore the opportunities for advancing community design and service provision in a more sustainable manner with a comprehensive, age-inclusive approach. We contend that older members of a community share many of the same physical, economic, and social needs as younger ones. For example, barrier-free buildings and sidewalk networks enhance the mobility and independence of seniors as well as children and their caregivers. Safe neighborhoods allow the young and old alike to participate in social activities and engage with other members of the community. Public transit, accessible parks, and senior-student mentoring are just a few of the other areas in which integrated service provision makes sense for local communities. Such a framework not only helps keep older adults engaged and active; it helps younger people grow towards independence.

Our examination focuses on policies and institutions in the United States, but we argue that the concept of multigenerational planning is universal. Both WHO's Age-Friendly Cities and UNICEF's Child Friendly-Cities have many elements in common (safety, housing, mobility, services, civic engagement) (WHO, 2007; UNICEF, 2004). By bringing the needs of children and older adults together, we can promote communities that better meet the needs of all ages.

At the same time, we argue that the quality of life and economic well being of the different generations are intertwined over the long term. In the United States, the increased percentage of retirees in our population of older adults threatens the national economic outlook as the economy loses workers with critical skills. With fewer workers, the growth in the U.S. gross domestic product is expected to slow (Myers, 2008). In addition, those fields directly linked to the increasing number of seniors, such as nurses and home health care aides, will rise with the population's average age, while the supply of people to fill those roles falls. Demographer Dowell Myers (2007) describes the social compact between generations: investing in young people, their health, and their education is an investment in the workforce, care providers, and taxpayers who will step into the societal roles vacated by aging baby boomers.

The shared reliance of different generations on one another is being recognized. A 2008 national survey of professional planners by Cornell University and the American Planning Association (APA) found that 90% agreed with the statement: "Communities that keep people for the whole life cycle (children, single adults, parents, elderly) are more vibrant." Sixty-four percent agreed that "the needs of families are similar to the needs of the elderly with regards to the physical environment (e.g. parks, transportation, affordable housing)" (Israel & Warner, 2008). The challenge is how to build a multigenerational planning approach. Analysis of the survey found that one of the biggest barriers to the creation of a family-friendly community is lack of understanding of what to do (Warner & Rukus, 2013).

Factors that limit multigenerational planning approaches include concerns over an inadequate tax base, economic development, and integration of services across age groups. Concerns about the tax burden that families with young children represent are balanced with recognition of the critical importance that families with young children represent for the economy—in both the short and long term. While 53% of planners responding to the APA survey agreed that, "Most families do not generate sufficient tax revenue to cover the cost of services they demand," 97% agreed with the statement that, "Families represent a valuable consumer population" (Israel & Warner, 2008). The USDA has tracked the contributions that families with children make, noting that, on average, families spend $222,360 from birth to age 17 on each child, and the majority (77%) of these expenditures (food, housing, transportation, education, health care) are spent in the local economy (Lino, 2010). These expenditures have multiplier effects that stimulate the broader economy, ensuring that critical services and jobs are available locally to benefit all. Childcare and other local services have been found to have multipliers higher than most typical targets for economic development, such as tourism (Kay, Pratt, & Warner, 2007). Economic developers are now shifting attention to local services as a critical aspect of local economic development (Warner & Liu, 2006; Warner & Zheng, 2013). The benefit of a vibrant local economy with a wide array of local services is that such communities are better places to grow up, to raise a family, and to age in place.

Effective economic development policy requires a multigenerational approach. While some communities seek to attract retirees for their disposable pension income and their support for local volunteer organizations (Glasgow & Brown, 2008), as seniors age, they require more services from the local community (para-transit, supported housing, meals on wheels) and costs rise. After age 85, a National Bureau of Economic Research analysis shows that the combined state and local expenditures on elders are higher on average than those incurred for school age children (Edwards, 2010). Many communities have chosen to invest in amenities that attract the creative class of young professionals without children. But when communities fail to offer the parks, childcare, and educational services that young families need, they risk losing these young professionals when they form families. Communities that invest in public amenities attractive to families, such as schools, safe streets, libraries, and parks, have been found to enjoy higher rates of economic growth (Reese, 2012).

Regarding services, age-segregated approaches to housing and services go against the wishes of many older adults. Research by AARP has shown that most aging Americans do not want to live in communities separate from younger people. A 2000 survey of adults older than age 55 found that 89% would like to stay in their current residence as long as possible (Bayer & Harper, 2000), and, increasingly, older adults are responsible for raising or helping to raise their grandchildren. In 2000, the U.S. Census found 5.8 million grandparents living in the same homes as their grandchildren, with 2.4 million of those older adults acting as heads of households. Most of those older adults were responsible for their grandchildren for five years or more. The trend is particularly strong among Latino households, which make up an increasing part of the population (Simmons & Dye, 2003). In 2008, 33% more Americans than in 1980 (or 15% of the population) lived in a family household that contained at least two adult generations or a grandparent, a sharp reversal from the previous four decades in which the number of multigenerational households was shrinking (Pew Research Center, 2010). The idea of separate physical places and separate services for old and young people does not match the reality of the way people live.

Our approach is particularly important as local governments face tight fiscal budgets. Public investment in seniors is primarily the responsibility of the federal government (Medicare and Social Security), while government service provision for young people is primarily the responsibility of state and local governments. A Brookings Institute study found that public expenditure per senior is almost two and a half times higher than public expenditure per child (Isaacs, 2009). Edwards (2010) estimates that public investment for young children up to age five is the lowest of any age group. Even in the best of times, local leaders have to balance the needs of one constituency against another. But by integrating services, it is often possible to meet the needs of both simultaneously. For example, communities can combine a senior center and childcare in one building. This is especially important in rural communities, which have lower per capita tax revenues, but higher per capita costs for services because of the sparse settlement patterns (Warner & Morken, 2013). One response, which works in urban, suburban, and rural communities, is shared services with the local school system (Vincent, Filardo, Allen, & Franklin, 2010). Such sharing may include the playground, the auditorium, the gym, the computer lab, or the school bus. Shared resources and programs can save money, promote intergenerational interaction, and enrich quality of life.

In all communities, the design of the built environment and the provision of services make more sense when they cater to more than one constituency. In this chapter, we explore three aspects of this approach to planning across generations. First, we look at some innovative ways that local governments, realizing the benefit of intergenerational planning and programming, have adopted important physical design guidelines and services in their communities that help both older adults and children. Second, in shaping these policies, we examine the important notion that participation by members of these age groups results in more robust and sustainable policies. Finally, by recognizing the ways that different generations cooperate, we build the case that older and younger citizens have a potentially powerful coalition to advocate for the services and community design standards that enable both to lead fuller, more independent, lives. Although this chapter focuses on U.S. municipalities, the theory can be translated to other nations and cultures.

BRIDGING GENERATIONAL BARRIERS

Children and their parents benefit from many of the same community components as older Americans. The similar needs of these groups include safe, walkable neighborhoods, a complete range of services nearby (childcare, senior centers, parks, food stores, health care, etc.), an opportunity for civic engagement, affordable and mixed-use housing, and adequate transportation options (Lynott et al., 2009). The lack of walkable neighborhoods is also a health concern for all ages. Half of adults older than age 50 do not get recommended levels of physical activity, and for those older than age 65, two-thirds do not get the exercise they need. As a result, nearly eight out of ten men and seven out of ten women older than age 60 are overweight, with about one-third considered obese (Flegal, Carroll, Ogden, & Curtin, 2010). At the same time, the rate of obesity in children aged two to five years has more than doubled since the late 1970s to more than 10.4%. For those aged six to 11, the rate of obesity tripled to 19.6% and for teenagers, obesity jumped from 5% to over 18% (Centers for Disease Control, 2010).

In addition to providing opportunities for exercise, walkable communities allow older adults and children without drivers' licenses to get out of their homes and into the community. Fully accessible neighborhoods help older adults, who are no longer comfortable behind the wheel or who do not drive any longer, to get around by themselves. Even without cars, they can go by foot or public transit to commercial or civic places where they can shop or interact with other community members. Young children and their caregivers also benefit from walkable communities and accessible public transit. The AARP has implemented two pilot programs in Richmond, Virginia, and Madison, Wisconsin, to increase activity by improving the physical environment in places where both students and older adults walk. The programs also conducted social marketing campaigns that raised awareness of the environmental barriers to walking and biking; conducted audits of 150 city blocks in Richmond and 30 residential streets in Madison; and crafted a plan for environmental and policy changes in each city. The organizers intentionally targeted programs and places that would help both older adults and school children. In Richmond, an intergenerational "Walk to School" event promoted the idea of getting relatives over 50 to walk children to school. The school district changed its policy to allow students to document their walking routes to school for future organized events. Many sidewalks, crosswalks, and intersections were repaired around town, especially near the schools and senior housing (Emery, Crump, & Hawkins, 2007).

A policy of "complete streets" formalizes a community's commitment to planning, designing, and building safe streets for users of all ages and abilities. By 2012, almost 500 local governments had adopted complete street policies (Seskin & Gordon-Koven, 2013). Typical elements of a complete street may include sidewalks, bicycle lanes (or wide shoulders), designated bus lanes, safe and accessible transit stops, and frequent safe crossing spots for pedestrians. Designs are customized for specific localities and regions, so rural roads will have a different set of requirements that make them accessible than urban ones (Smith, Reed, & Baker, 2010).

There are limitations to what a municipality can do to reshape its built environment to the benefit of multiple generations. The physical redesign can take years to implement and may be beyond the capacity of local governments. An enabling redesign may be rendered more difficult by longstanding development patterns, such as those typically found in suburban and rural areas. The gap between a fully enabled environment for residents and the ability of a municipality physically to make such redesigns is usually filled by the integrated provision of services (Warner & Morken, 2013). For example, in neighborhoods of New York City that lack neighborhood grocery stores, seniors have difficulty accessing shops and civic amenities. Redesigning the built environment is a longer-term solution, but in the short term, a service overlay can make the difference. The New York City schools teamed up with the Department of Aging to transport older New Yorkers from senior centers to supermarkets, as well as to museums, parks, and other public places. The program uses school buses when they are not moving children. For older adults, the trips are free, and the city has found a way to make more efficient use of the large capital investment made in the buses. The multigenerational bus strategy took planning and coordination between two New York City bureaucracies with distinct missions and target populations (Akam, 2009). This kind of transportation strategy might be even more cost-effective in many suburban and rural places where public transit is poor or non-existent, but school buses are plentiful and only used a fraction of each day.

School buildings are other assets that may be underutilized. During the day, they might be full of students. On weekends, in the evenings, and over the summer, they usually sit empty. School buildings are often located in neighborhoods where they can create strong communities when they act as venues for neighbors to build relationships and a cohesive community. Broader benefits can accrue to the community, especially for older adults who could use the building for meals or recreation. They could also assist with instruction and other school activities (Council of Educational Facility Planners International, 2004). What is needed is sensitive architectural and programmatic design so that the distinct requirements of each age group can be met and liability and security concerns addressed (Dunn, 2013). Community schools, which can serve as one-stop shops for a range of community services, are becoming more common. For example, in 2001, Cincinnati's Board of Education redeveloped all schools in the district as community learning centers. The $1 billion effort engaged neighborhoods in the planning, implementation, and governance of each school with services beyond education and based upon local needs. The Oyler School, in one of the poorest neighborhoods, offers a full-service health center, access to vision and dental clinics, and three full-time psychologists. There also are evening meals for students and their families, food to tide over the neediest children on the weekends, college and career planning, varied after-school programming, and more (Morken & Baran-Rees, 2012).

Intergenerational educational programs can foster positive outcomes for all participants. In Ithaca, New York, a local Head Start program is permanently housed at a retirement community. Each week, the older adults work with preschoolers on a variety of activities such as reading, singing, and crafts. In Jenks, Oklahoma, the public school district buses 60 kindergarten and prekindergarten students to classrooms in the Grace Living Center, a nursing home in this Tulsa suburb. The nursing home built the two classrooms in their facility, which they lease back to the school district for a dollar per year. In them, 170 older adults mentor the children in their academic and social development. Tracking by the school district has found that 10% fewer children who attend preschool and kindergarten at the nursing home need reading interventions when they enter the first grade (Morehouse, 2008). Formal studies of such structured interactions between young children and older adults show that children become more helpful, empathize with older people, and develop better self-control as a result (Femia, Zarit, Blair, Jarrott, & Bruno, 2008).

There are a number of barriers to such intergenerational programming and joint use, but these can be overcome with careful planning (Morken & Baran-Rees, 2012). Many federal and state funding streams require funds to be used only for one target population (children or elders, not both). Liability is also a major concern, particularly for school administrators, who worry about injuries incurred on school property; this proved to be a perceived barrier even if state law offered protection to local school districts (Spengler, Connaughton, & Maddock, 2011). The safety of older or younger people, when the populations are mixed, is a common security concern that hinders intergenerational programming (Meehan & Sanchez, 2012; Pynoos & Parrott, 1996). Insurance companies can be hesitant to cover, for example, paratransit companies that transport mixed populations out of fear that drivers are not fully trained to handle the diverse service needs (Farber, 2008). However, these liability concerns can be overcome, as shown by many communities that promote shared services with schools. NPLAN (2010) collects data on state laws addressing how to overcome these liability issues.

PARTICIPATION IS KEY

The involvement of stakeholders in planning for physical changes or service provision that meet the needs of all generations is important. Collaborative planning involves broadening perspectives (Innes & Booher, 2003; Forester, 1999). For example, members of different age groups use public space in somewhat different ways and at different times; thus, changes in design can affect young and old differently (Peace, 2013). Affected citizens bring more than their local knowledge to a process. Their involvement also brings ingenuity, capacity, and political accountability (Fung, 2008). These points are particularly true for older adults who have decades of local knowledge and are often seasoned in problem solving. Youth bring a fresh perspective, as well as their understanding of what is important to their peers. Participation is core to both the WHO and UNICEF approaches to building age- and child-friendly cities.

Atlanta has taken this message to heart in the development of its Lifelong Communities initiative. Here, participation of all community members is important in determining how to address the unique challenges of redesigning a sprawling suburban landscape better to meet the needs of both older adults and children (Morken, 2012). Lifelong Communities surveyed 1,500 older adults and professionals from a wide variety of fields and hosted a nine-day urban design charrette to solicit older adults' views and promote a cross-sectoral response to improving the community for older adults (Duany & Plater-Zyberk, 2009; Keyes et al., 2013). Age-Friendly NYC has involved thousands of seniors in surveys and focus groups in its efforts to get city agencies and local businesses to become more sensitive to the needs of older adults (Finkelstein, Garcia, Netherland, & Walker, 2008).

As part of the 2020 Community Plan on Aging in the Charlottesville, Virginia, area, planners decided intentionally to be age-inclusive. High school students were recruited as members of the planning committee. They acted as ambassadors to other young people through focus groups and student surveys. In the end, the high school students wrote a chapter of the plan titled "Strengthening Intergenerational Connections" with recommendations that included recruiting students as healthcare workers; encouraging alternative transportation options; promoting intergenerational volunteering to bring together older adults and youth in meaningful service; and educating youth on the need for lifelong financial planning. One outcome of this intergenerational planning was a program that recruited more than 20 older adults to volunteer in seven elementary schools to tutor reading, math, and languages, as well as to provide library and landscaping assistance (Jefferson Area Board for Aging, 2003).

Kaplan and colleagues (2004) describe a series of participatory action-research events that increased the participation of different age groups in a community-visioning process. The process uses various media, such as murals, models, photographs, theatrical displays, and other communications, to engage youth and older adults together. The process, known as a FUTURES festival, brought young and old together in Kaneohe, Hawaii, to work out conflicting visions for a local park. The intergenerational participation resulted in the design of a "Park for All Ages" that included areas for skateboarding, shuffleboard, picnic areas, and a Braille trail (Kaplan, 2001).

Participation by different age groups builds good citizens, as well as community engagement. Research shows a positive association between engaging in civic activities and better health in later life (Hinterlong, Morrow-Howell, & Rozario, 2007). Participation provides the opportunity to give back to the community. The younger end of the spectrum benefits as well. Involving children in public planning processes can foster local knowledge and environmental responsibility and can promote personal development and citizenship (Frank, 2006). Research has found that involving families and seniors in the planning process leads to more age-sensitive planning and services focused on the needs of children and seniors (Warner, Homsy, & Morken, 2013; Warner & Rukus, 2013). Participation in planning is especially important in overcoming limited resources or more challenging environments such as those found in rural and suburban communities.

INTERGENERATIONAL COALITIONS

The links between the generations, as outlined in this chapter, may seem obvious. Indeed, according to the 2008 survey by Cornell University and the American Planning Association, nine out of ten planners understand that communities populated by people of every age bracket are more vibrant, and about two-thirds recognize the connection between the needs of older adults and those of families with children regarding physical design (Israel & Warner, 2008). The problem, the survey found, is translating this understanding of multigenerational communities into action on the ground. A regression analysis of the 900 planners responding to the survey found that family participation and site planning and zoning were the most significant drivers of action. Resistance was primarily driven by ignorance of where to begin. Focusing on details of site planning and zoning and family participation in planning can overcome these challenges (Warner & Rukus, 2013).

Evidence of the strength of such a coalition can be found in Gaylord, Michigan. After being twice thwarted in its plans to construct a new school building, the school board in this rural community tried a new approach by incorporating broader community needs and desires into the project. The board's revised strategy was a facility that includes childcare, a community health care clinic, senior activities, a performing arts center, and a gymnasium and classrooms designed for shared community use. The new $25 million bond proposal fostered a broad sense of community investment in its school and eventually gained support of Gaylord residents. The school gave Gaylord its first performing arts center, which includes practice space for band, chorus, and other ensembles, and, with its many uses, has become a place where students and other community members regularly interact. Similarly, in Emeryville, California, a new stakeholder group formed in the wake of a fiscal crisis in the education system to push through a new tax critical to the schools. The redeveloped Emeryville Center of Community Life, jointly developed by the school district and the city, brings together education, recreation, and a range of services in the school building (Morken & Baran-Rees, 2012).

In-depth stakeholder analyses still need to be undertaken to better conceptualize the policymaking process across generations. However, it seems clear that advocates for older adults and children share many core values and can find many avenues of coordinated action to push for change. Nationally, several coalitions of advocates have already staked out positions in the policy landscape. AARP, which is already a politically powerful advocate for older adults, has taken up issues of children, especially because many older adults care for their grandchildren. The organization has a seat on the executive committee of the National Complete Streets Coalition. AARP is currently developing livability indicators with attention to the mobility and housing needs of older adults but recognizing the commonality of interests with young children and their families. An Aging White Paper, presented to the American Planning Association at its national conference in 2013, called for "quality growth to support multi-generational communities," in addressing the challenge planners face in meeting the needs of an aging population (Blanton & Bowen, 2013, p. 2).

Within municipalities, especially smaller cities, suburban towns, and rural communities, older adults can similarly play leadership roles in pushing for change that includes support for children and their families. At the local level, residents can direct their energies toward redesign of the built environment and reshaping programs to benefit both seniors and children. Given that many local leaders are driven by a desire to control property taxes, planning for all generations offers a way to translate these values into cost-efficient policies that improve quality of life for all segments of the community.

An aging society creates new stressors and new opportunities for community building. By building political coalitions across young and old, conflicts between generations over resources can be reduced, and communities can ensure that everyone's needs are met. Common vision can be created among the young, the old, and the sandwich generation of caretakers if attention is paid to everyone's needs for inclusive design and the potential for shared services. The younger generation is even more diverse than the baby boomers, so care must be given to structure programs that meet the needs of diverse ethnic and income groups. Myers (2007) calls this the new social compact between immigrants and boomers.

PLANNING ACROSS GENERATIONS

Multigenerational planning for lively and healthy communities specifically addresses the needs of all age groups and people of all abilities. The American Planning Association acknowledges the critical importance of the built environment, planning, service delivery, and building political coalitions to address the needs across all generations (Ghazeleh, Greenhouse, Homsy, & Warner, 2011). As communities experience dramatic shifts in demographics, the demand for age inclusive communities will intensify. The pressures of an aging population make the recognition of shared values crucial. Society in general and local governments in particular can benefit by addressing the needs of seniors, children, and caregivers together. Communities need to be designed to address the needs of all ages. The promise of multigenerational planning is that it enhances the independence of children and elders and reduces dependence on caregivers, thus enhancing the capacity of all residents.

In this chapter, we have shown that a multigenerational planning approach is an effective way to build better communities for aging in place. Such a lifecycle approach has economic development benefits for the entire community and enhances service effectiveness by promoting synergies across service silos. This reduces costs to local governments and taxpayers. Additionally, by encouraging broad participation of all members of the community, including the youngest and oldest members, a multigenerational approach builds a shared vision and the political support needed for change. Multigenerational communities are communities for all. When the needs of both the youngest and the oldest residents are recognized in planning, service delivery, and economic development, and when their civic participation is encouraged, communities have the ability to create sustainable policies and programs that benefit everyone.

Acknowledgments

This research was made possible in part with support from the U.S. Department of Agriculture, National Institute for Food and Agriculture research grant #2011-68006-30793 for the Planning Across Generations Project. More information can be found at www.mildredwarner.org.

REFERENCES

Akam, S. (2009, April 26). Bloomberg plan uses school buses to shuttle older residents. *The New York Times*. Available from http://www.nytimes.com/2009/08/26/nyregion/26elders.html

Bayer, A.-H., & Harper, L. (2000). *Fixing to stay: A national survey of housing and home modification issues*. Washington, DC: AARP Independent Living Program. Available from http://assets.aarp.org/rgcenter/il/home_mod.pdf

Blanton, W., & Bowen, C. (2013). *APA legislative & policy committee policy paper: Planning for an aging society*. Presented to 2013 National Delegate Assembly, American Planning Association, Chicago, IL.

Centers for Disease Control. (2010). Childhood overweight and obesity. *Obesity and Overweight for Professionals: Childhood | DNPAO | CDC*. Available from http://www.cdc.gov/obesity/childhood/index.html

Council of Educational Facility Planners International. (2004). *Schools for successful communities: An element of smart growth.*

Duany, A., & Plater-Zyberk, E. (2009). *Lifelong communities: A regional guide to growth and longevity*. Atlanta, GA: Atlanta Regional Commission. Available from http://www.atlantaregional.com/File%20Library/Aging/ag_llc_regional_guide.pdf

Dunn, D. (2013). Making it work: Designing neighborhood schools for the entire community. Ithaca, NY: Cornell University. Available from http://cms.mildredwarner.org/p/178

Edwards, R. (2010). *Forecasting government revenue and expenditure in the U.S. using data on age-specific utilization* (Working Paper # WP10–01, National Transfer Accounts Project). Available from http://qcpages.qc.cuny.edu/~redwards/Papers/edwards-forecasting-0210.pdf

Emery, J., Crump, C., & Hawkins, M. (2007). Formative evaluation of AARP's active for life campaign to improve walking and bicycling environments in two cities. *Health Promotion Practice*, *8*(4), 403–414. doi:10.1177/1524839906292179

Farber, N. (2008). *Insurance challenges for paratransit*. Denver, CO: National Conference of State Legislatures. Available from http://www.ncsl.org/print/transportation/insurancebrief1008.pdf

Femia, E.E., Zarit, S.H., Blair, C., Jarrott, S.E., & Bruno, K. (2008). Intergenerational preschool experiences and the young child: Potential benefits to development. *Early Childhood Research Quarterly*, *23*(2), 272–287.

Finkelstein, R., Garcia, A., Netherland, J., & Walker, J. (2008). *Toward an age-friendly New York City: A findings report*. New York: The New York Academy of Medicine. Available from http://www.nyam.org/news/docs/AgeFriendly.pdf

Flegal, K.M., Carroll, M.D., Ogden, C.L., & Curtin, L.R. (2010). Prevalence and trends in obesity among US adults, 1999–2008. *JAMA, 303*(3), 235.

Forester, J. (1999). *The deliberative practitioner: Encouraging participatory planning processes*. Cambridge, MA: The MIT Press.

Frank, K.I. (2006). The potential of youth participation in planning. *Journal of Planning Literature, 20*(4), 351–371. doi:10.1177/0885412205286016

Fung, A. (2008). Citizen participation in government innovations. In S. Borins (Ed.), *Innovations in government: Research, recognition, and replication* (pp. 52–70). Washington, DC: Brookings Institution Press.

Ghazaleh, R.A., Greenhouse, E., Homsy, G., & Warner, M. E. (2011). *Using smart growth and universal design to link the needs of children and the aging population* (APA Briefing Paper). Washington, DC: American Planning Association. Available from http://www.planning.org/research/family/briefingpapers/multigenerational.htm

Glasgow, N., & Brown, D. (2008). *Grey gold: Do older in-migrants benefit rural communities?* (Paper No. 58). Durham, NH: Carsey Institute, University of New Hampshire. Available from http://scholars.unh.edu/carsey/58

Hinterlong, J.E., Morrow-Howell, N., & Rozario, P.A. (2007). Productive engagement and late life physical and mental health: Findings from a nationally representative panel study. *Research on Aging, 29*(4), 348–370. doi:10.1177/0164027507300806

Innes, J.E., & Booher, D.E. (2003). Collaborative policymaking: Governance through dialogue. In M. Hajer & H. Wagenaar (Eds.), *Deliberative policy analysis: Understanding governance in the network society* (pp. 33–59). Cambridge, UK: Cambridge University Press.

Isaacs, J. (2009). *How much do we spend on children and the elderly*. Economic Studies Report. Washington, DC: The Brookings Institution. Available from www.brookings.edu/reports/2009/~/media/68F7781F1612493E8AD0F419F7A35E65.pdf

Israel, E., & Warner, M. (2008). Planning for family friendly communities. *PAS Memo*. Washington, DC: American Planning Association. Available from http://www.planning.org/pas/memo/open/nov2008/index.htm

Jefferson Area Board for Aging. (2003). *2020 community plan on aging*. Charlottesville, VA: Jefferson Area Board for Aging. Available from www.jabacares.org/page/full/2020-plan/

Kaplan, M. (2001). *The Futures Festival: An intergenerational approach to community participation* (No. CAT UI364 1M11/01ps4475). University Park, PA: Penn State College of Agricultural Services—Cooperative Extension.

Kaplan, M., Higdon, F., Crago, N., & Robbins, L. (2004). Futures Festivals. *Journal of Intergenerational Relationships, 2*(3–4), 119–146. doi:10.1300/J194v02n03_10

Kay, D.L., Pratt, J.E., & Warner, M.E. (2007). Role of services in regional economy growth. *Growth and Change, 38*(3), 419–442.

Keyes, L., Phillips, D., Sterling, E., Manegdeg, T., Kelly, M., Trimble, G., & Mayerik C. (2014). Transforming the way we live together—a model to move communities from policy to implementation. *Journal of Aging & Social Policy, 26*(1&2), 117–130. doi:10.1080/08959420.2014.854646

Lino, M. (2010). *Expenditures on children by families, 2009* (Misc. Pub. No. 1528–2009). Washington, DC: U.S. Department of Agriculture, Center for Nutrition Policy and Promotion. Available from www.cnpp.usda.gov/publications/crc/crc2009.pdf

Lynott, J., Taylor, A., Twaddell, H., Haase, J., Nelson, K., McCann, B., & Stollof, E.R. (2009). *Planning for complete streets for an aging America* (No. 2009–02). Washington, DC: AARP Public Policy Institute.

Meehan, S., & Sanchez, T. (2012). *Coordinating school and public transportation: Assessing the opportunities in New Jersey.* New Brunswick, NJ: Alan M. Voorhees Transportation Center, Rutgers University.

Morehouse, L. (2008, January 28). Senior citizens help young children with reading—and relationships. *Edutopia.* Available from http://www.edutopia.org/grace-learning-center-prekindergarten-community

Morken, L. (2012). *Cities plan for the aging population: Lessons from New York City and Atlanta.* Ithaca, NY: Department of City and Regional Planning, Cornell University. Available from http://cms.mildredwarner.org/p/134

Morken, L. J., & Baran-Rees, R. (2012). *Joint use: School-community collaboration.* Ithaca, NY: Department of City and Regional Planning, Cornell University. Available from http://cms.mildredwarner.org/p/147

Myers, D. (2007). *Immigrants and boomers: Forging a new social contract for the future of America.* New York: Russell Sage Foundation.

Myers, D. (2008). *Old promises and new blood: How immigration reform can help America prosper in the face of baby boomer retirement.* Alexandria, VA: Reform Institute.

National Policy and Legal Analysis Network (NPLAN). (2010). *Fifty-state scan of laws addressing community use of schools.* Available from http://changelabsolutions.org/sites/changelabsolutions.org/files/documents/JU_StateSurvey_FINAL_2010.03.19.pdf

Peace, S. (2013). Social interactions in public spaces and places: A conceptual overview. In G. Rowles & M. Bernard (Eds.), *Environmental gerontology: Making meaningful places in old age* (pp. 25–49). New York: Springer Publishing Company.

Pew Research Center. (2010). *The return of the multi-generational family household.* Washington, DC: Pew Research Center.

Pynoos, J., & Parrott, T. (1996). The politics of mixing older persons and younger persons with disabilities in federally assisted housing. *The Gerontologist, 36*(4), 518–529. doi:10.1093/geront/36.4.518

Reese, L. (2012). Creative class or procreative class: Implications for local economic development. *Theoretical and Empirical Researches in Urban Management,* 7(1), 5–26.

Seskin, S., & Gordon-Koven, L. (2013). *The best complete streets policies of 2012.* Washington, DC: Smart Growth America.

Simmons, T., & Dye, J. L. (2003). *Grandparents living with grandchildren: 2000* (No. C2KBR-31). Washington, DC: U.S. Census Bureau.

Smith, R., Reed, S., & Baker, S. (2010, August). Street design: Part 1—complete streets. *Public Roads,* 74(1). Washington, DC: U.S. Department of Transportation. Available from http://www.fhwa.dot.gov/publications/publicroads/10julaug/03.cfm

Spengler, J. O., Connaughton, D. P., & Maddock, J. E. (2011). Liability concerns and shared use of school recreational facilities in underserved communities. *American Journal of Preventive Medicine, 41*(4), 415–420. doi:10.1016/j.amepre.2011.06.031

UNICEF. (2004). *Building child friendly cities: A framework for action.* Florence, Italy: UNICEF Innocenti Research Centre. Available from http://www.unicef-irc.org/publications/416

Vincent, J., Filardo, M., Allen, M., & Franklin, J. (2010). *Joint use of public schools: A framework for a new social contract.* Washington, DC: 21st Century School Fund and Center for Cities and Schools.

Warner, M. E., Homsy, G., & Morken, L. (2013). *Planning for aging in place: Stimulating a market and government response.* Ithaca, NY: Department of City and Regional Planning. Available from http://www.mildredwarner.org

Warner, M. E., & Liu, Z. (2006). The importance of child care in economic development: A comparative analysis of regional economic linkage. *Economic Development Quarterly, 20*(1), 97–103.

Warner, M. E., & Morken, L. (2013). Building child and age-friendly communities in tight fiscal times. In *The municipal year book 2013* (pp 47–56). Washington, DC: International City County Management Association. Available from http://cms.mildredwarner.org/p/170

Warner, M. E., & Rukus, J. (2013). Planners' role in creating family friendly communities: Action, participation and resistance. *Journal of Urban Affairs*, *35*(5), 627–644. doi:10.1111/juaf.12014

Warner, M. E., & Zheng, L. (2013). Business incentive adoption in the recession. *Economic Development Quarterly*, *27*(2), 90–101.

World Health Organization (WHO). (2007). *Global age-friendly cities: A guide*. Geneva, Switzerland: World Health Organization. Available from http://www.who.int/kobe_centre/publications/age_friendly_cities_guide/en/index.html

14 Age-Friendly Community Policy Innovation

Complete Streets Implementation in Louisiana, United States

Billy Fields and Jason Tudor

INTRODUCTION

Increasingly, policymakers, researchers, and practitioners are examining the role that safe, walkable places play in encouraging active, healthy communities for older adults. Recent research points to the importance of accessible places with multiple transportation options as a key factor in promoting physical activity, overall health, and the ability to age in place for older citizens (Kerr, Rosenberg, & Frank, 2012; Wahl, Iwarsson, & Oswald, 2012).

Despite the benefits of walkable, accessible places for older adults, car-dominated landscapes are the norm in the United States, with walking rates for older Americans far lower than in many other industrialized countries. Pucher and Dijkstra (2003) point out that walking and cycling are used for only 6% of all trips in the United States for Americans over 65, compared to close to half of all trips for Dutch and German citizens over 75. While the United States ranks towards the bottom of measures of walking by older citizens, the problem is shared by other countries like Australia and Canada, where walking and cycling rates remain low (Bassett, Pucher, Buehler, Thompson, & Crouter, 2008).

While differences in culture from nation to nation can explain some of the differential in walking rates for older adults, differences in transportation policy play a pivotal role in creating transportation choices (Lynott et al., 2009). Most communities in the United States are still planned almost exclusively for the automobile, with a design emphasis on maximizing vehicle speed and lane number and width (Wells, 2013). The resulting landscape limits transportation choices for older adults by making pedestrian street crossings more difficult. This stems from the policies and practices of transportation service organizations, such as departments of transportation that focus almost exclusively on auto-transportation service delivery without considering other system such as pedestrians, transit users, and bicyclists.

This chapter analyzes the opportunities and barriers associated with changing the entrenched policy framework of departments of transportation to include more age-friendly community components that facilitate walking and access to transit. The political and administrative challenges are analyzed through a case study of the passage and implementation of a highly-ranked complete streets policy in the state of Louisiana in the United States (Seskin & Gordon-Koven, 2013). Complete streets policies are emerging as a key policy innovation designed to create safe streets for users of all ages and abilities. Analysis of the complete streets policy process through an overview of key events and qualitative analysis of key stakeholder interviews provides an important window into the institutional framework (Healey, 2004) that surrounds transportation policymaking and shows how coalition building and pragmatic advocacy work can build political momentum for more age-friendly policies.

The chapter is organized into three sections. First, the connections between neighborhood street design and age-friendly communities are analyzed. This is followed by an overview of the key institutional structures that control the design and management of streets in the United States. Finally, the Louisiana complete streets policy process is analyzed.

COMPLETE STREETS AND AGE-FRIENDLY COMMUNITIES

The focus of complete streets policies is to help provide, as the National Complete Streets Coalition describes, "safe access to destinations for everyone, regardless of age, ability, income, ethnicity, or how they travel" (Seskin & Gordon-Koven, 2013, p. 1). Complete streets policy adoption has grown significantly over the last 10 years, with 525 policies currently adopted across the United States (Snyder, 2013). The movement is expanding internationally as well, with growing interest from communities in Canada (Complete Streets for Canada, 2013) and Australia (Burgess et al., 2010).

One of the key demographic groups that can benefit from complete streets policies is older adults. While the older adult demographic population is increasing significantly (He, Sengupta, Velkoff, & DeBarros, 2005), few structural changes from a transportation perspective have been made to help residents age in place. Recent surveys of older adults show the mismatch between the desire and the ability to age safely in place. While 90% of adults want to age in the home where they currently live (Farber, 2011), 50% of older adults report that they cannot cross roads close to their homes safely while walking (Lynott et al., 2009). This perception is matched by crash statistics, which show that 22% of all pedestrian fatalities were older adults over age 65 (Ernst, 2011).

The absence of safe walking opportunities has significant implications for older adults since the lack of access to a car due to financial or health reasons increases with age. Bailey (2004), for example, surveyed older adults in the United States and found that 21% of the older adult population over 65 does not drive. She argues that the result is that, "Older non-drivers have a decreased ability to participate in the community and the economy" (Bailey, 2004, p. 1).

Complete streets policies coupled with access to transit offer a much wider set of mobility choices that can increase the opportunities to age in place. At the heart of this approach is fostering walkable communities. Lehning (2012), in a summary of recent research on the benefits of walkable communities for older adults, notes that these types of communities "increase the life space" of older adults, improve access to shopping opportunities, improve social capital of residents, improve levels of physical activity, and help maintain levels of instrumental activities for daily living (p. 346).

From a transportation design perspective, the changes needed to foster walkable communities with connections to transit are surprisingly simple and cost-effective. Kerr et al. (2012), in a review of key built environmental features and healthy aging, note that communities with well-maintained sidewalks, crosswalks at signalized intersections, extended walk times for signals, decreased crossing distances through use of medians, traffic calming, and speed management are effective at encouraging older adult walking. These are essentially the changes encouraged by complete streets policies.

U.S. TRANSPORTATION POLICY AND AGE-FRIENDLY COMMUNITIES: ACTORS AND INSTITUTIONS

While the complete streets changes proposed above are fairly straightforward from a design perspective, they represent a significant change in priorities for transportation organizations. One way better to understand and identify potential implementation barriers to more age-friendly policies is to examine the network of actors involved in transportation policy decision-making and implementation. This institutional approach focuses on the analysis of changing agency practices, policies, and actors in relation to broad structures and the governance culture that frames decisions (Healey, 2004). This section provides an overview of the key institutions and actors involved in U.S. transportation policy.

Current policies in transportation departments across the United States focus almost entirely on the safe movement of automobiles in the transportation system (Johnson & White, 2010). This innocuous and necessary task has, however, often been seen as the sole mission of transportation departments, leaving walking, cycling, and transit out of the planning and design process (Lynott et al., 2009). Complete streets policies represent a challenge to these routine practices as they seek to create a more holistic design approach focused on creating safe transportation for users of all ages and abilities.

Changing the entrenched organizational behavior of transportation departments is, however, a significant challenge (Fields & Hull, 2013). Over time, routine practices associated with auto-oriented engineering have been established. These practices define the dimensions of the transportation policy area, key actors involved in the process, and the scope of action that agencies take in addressing that area. Moving towards accommodating new actors and issues requires changing these agency practices and moving beyond this administrative inertia (Brown, 2008; Downey, 2007).

Currently, transportation policy in the United States is set through a complicated network of federal, state, and local decision-making designed to provide flexible responses to local conditions (Downey, 2007). While much of the funding for transportation projects originates at the federal level through large transportation funding bills, federal guidance through the U.S. Federal Highway Administration (FHWA) on project selection is often minimal, with deference shown to state and local governments. State departments of transportation are key actors as they act as controlling agents for project funding and set rules for design and management of projects on state highways. Their professional organization, the American Association of State Highway and Transportation Officials (AASHTO), is a key actor in rulemaking and federal lobbying. Finally, local departments of public works coordinate with regional Metropolitan Planning Organizations (MPOs) to select and fund projects in local communities.

Decisions about the design of roadways are made by transportation engineers from these multiple levels of government working from a set of professional design manuals (Griffin, 2011). These manuals define the standard operating procedure of these agencies and are foundational documents that transportation engineers rely on to make design decisions ranging from lane width to permitted speed (Lynott et al., 2009). These manuals are traditionally adjusted through a slow, conservative process of research and evaluation.

Despite a broad focus on multimodal transportation ushered in by the 1991 federal transportation bill ISTEA (Intermodal Surface Transportation Efficiency Act), most of the focus of these design manuals continues to be on creating designs that safely maximize auto capacity. Speed and volume maximization are the key considerations in the design process as traffic engineers work to model traffic throughput on roadways (Figure 14.1).

Complete streets policies seek to alter this standard design process by creating a more holistic design problem for engineers that includes the needs not only of auto users, but of pedestrians and cyclists as well. Instead of focusing on a single design problem of maximizing auto use, the complete streets approach seeks to balance access for multiple modes. This balanced approach does not have a single answer, but instead is more of a process of building solutions acceptable to the broader community (Figure 14.2).

While institutional actors and professional expectations were clearly set under the highway traffic engineering model, the complete streets approach with new design users presents a significant challenge to many traffic engineers who have spent their entire careers working under the paradigm of maximizing auto flow (Fields & Hull, 2013). Complete streets policy adoption and implementation occurs within this framework of established administrative practices, colliding with new policy actors such as public health,

Figure 14.1 Auto-Oriented Street Example

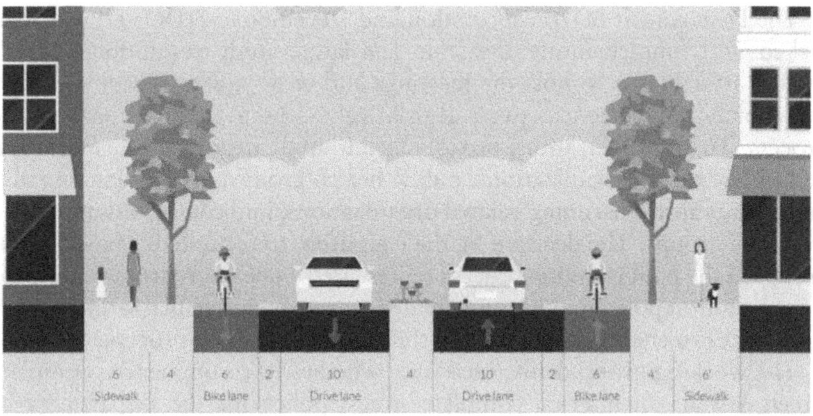

Figure 14.2 Complete Streets Design Example

complete streets, bicycles and pedestrians, smart growth, and age-friendly advocates seeking to establish more holistic standards, designs, and federal funding flows (Handy et al., 2009; Mapes, 2009). The heart of the complete streets policy challenge is coordinating this transition in administrative practices and governance structures to produce projects that allow multiple users of all ages and abilities safely to utilize the transportation system.

RECOGNIZING THE NEED: BACKGROUND TO THE LOUISIANA CASE

These broad structuring forces form an important background for understanding the challenges and opportunities in the adoption and implementation of the Louisiana complete streets policy. Although there are many issues that make aging in place more difficult for older adults in Louisiana, including access to primary health care, coordination of human services transportation, and affordable and accessible housing, access to safe neighborhood transportation is a significant need. Louisiana ranks third in the number of pedestrians of all ages injured in motor vehicle crashes in the United States, with 90 older adult pedestrians killed between 2000 and 2007 (Transportation for America, 2011).

Because of the significant pedestrian crash problem, AARP Louisiana, along with other active transportation advocacy groups, worked with the Louisiana legislature to draft a study resolution in 2009 to convene a workgroup to examine the need for a complete streets policy. The draft resolution encouraged collaboration between multiple groups to decide what a complete streets policy would look like in Louisiana, but prescribed no specific outcome. This resolution was eventually approved by the Louisiana state legislature in 2010.

The Department of Transportation and Development (DOTD) was then tasked with implementing the state legislative study resolution. DOTD funded a facilitator to host the meetings and write a final report. Over 35 stakeholders and agencies were legislatively chosen to participate in the process. These organizations broadly fit into five categories: transportation and public safety organizations, public health groups, urban planning and design organizations, aging-related organizations, and education/safe routes to school groups. The decision by the legislature to include this broad set of groups in the policymaking process represented a departure from traditional transportation policymaking and created a more inclusive platform for multiple voices to engage in the transportation policymaking process.

The working group deliberated over whether to recommend a complete streets law or an internal complete streets policy for the Department of Transportation and Development. The final recommendation by the workgroup was for the Department of Transportation and Development to adopt an internal policy. In 2010, the DOTD adopted an internal complete streets policy to guide project design and funding. The policy was ranked as the second best state policy in the country by the National Complete Streets Coalition based on key policy metrics (Seskin & Gordon-Koven, 2013).

A great policy is, however, only as good as its implementation. Although DOTD created complete streets checklists for project design, helping to begin to institutionalize complete streets implementation, many advocates have been dissatisfied with implemented projects that seem to lack basic complete streets principles. To ensure the implementation of the policy, AARP and other partners worked with legislators to submit a second resolution in 2012 to reconvene the workgroup to study the progress of implementation. In 2013, the DOTD reconvened the work group without a legislative resolution to continue assessing implementation of the policy. At this point, the complete streets policy implementation in Louisiana appears to represent more of a long-term process of change rather than a completed set of actions. Some communities in Louisiana, such as New Orleans, have created systematic approaches to ensure complete streets implementation. Many others communities, however, have utilized a much less structured, ad hoc approach to complete streets implementation.

RESEARCH METHODS

To understand the deeper meaning of the institutional process described above, a series of nine in-depth, qualitative interviews was conducted with key stakeholders to analyze the opportunities and potential barriers to the adoption and implementation of the Louisiana complete streets policy. The qualitative approach used in this chapter follows Miles and Huberman's (1994) basic framework of interviews, followed by data reduction and coding of themes, data display of the themes, and analysis. This approach allows key themes to emerge from stakeholder experiences in a structured format of iterative evaluation and reexamination. Quotes are used as "data" to outline key themes of the case in the words of the respondents. These quotes are attributed to respondents only through broad professional affiliation to preserve anonymity. As this is one of the first evaluations of complete streets policy adoption and implementation, the qualitative approach of building key themes provides an important base for future research (Johnson & White, 2010).

The interview questions were developed based on the application of the age-friendly and institutional literatures surveyed above to the complete streets policy sphere. The semi-structured questions were developed around the key areas of policy adoption, implementation, policy impact, and importance of age-friendly concepts to the policy sphere. A broad set of stakeholders, including Department of Transportation and Development (DOTD) and MPO employees, urban planners, nonprofit members, and public health representatives, were interviewed for the study. This broad network of stakeholders provides the advantage of capturing the full spectrum of engaged viewpoints on the policy process. More in-depth analysis of specific professional categories such as the DOTD employees may be an important area of future research in order to capture professional sector viewpoints. Despite this limitation, this analysis uncovered a set of key themes that recurred across the broad network of stakeholders.

A New Paradigm: Meaning of Complete Streets

A central theme that recurred throughout the interviews was the importance of complete streets as a new and important paradigm for transportation planning. Prior to the complete streets approach, the transportation model at the Louisiana DOTD was described by one of the non-profit respondents as a "highway focused department" that provided little "attention to walking and biking and design for that on state roadways."

The complete streets approach, by comparison, was seen as offering a more holistic, age-friendly system of transportation choices to all vulnerable road users. One of the public health respondents pointed out that the complete streets approach was more than just designing for "children within school zones or elderly people around assisted living homes" but instead was about focusing on "vulnerable users whenever we do design."

This change in approach involves taking seriously the needs of the approximately 30% of the overall population that does not or cannot drive to create a full transportation system that provides choices for all user groups (Millar & Sundquist, 2012). One of the MPO respondents argued that complete streets was central to providing "choices . . . to the traveling public and there are ways to provide those choices without huge costs and without radical changes in the way we are doing things." Instead of a set of "radical changes," the complete streets approach was seen as offering a realistic way forward to meet the needs of multiple users through targeted design changes.

The complete streets concept was also seen as an important unifying framework for explaining why these design changes needed to be included in projects. One of the public health respondents who had worked for years to get more holistic designs incorporated into projects argued that the "culture of talking around complete streets simplifies my work a lot. Complete streets has become a rally cry around what we want in our communities. Now I don't need to talk about bike and ped and transit. It all fits. The term has been around for 10 years, but we never used it and suddenly it became OK to use it." The respondent went on to argue that complete streets has provided a "validation of what we're doing. Complete streets . . . becomes a cultural shift."

While this "cultural shift" presents an important platform for engineers to redesign public roads for all users, several respondents pointed out the limitations of this traffic engineering approach to improving accessibility for shopping and other daily tasks. One of the consultant respondents noted that, "Something that's been a worry of mine is that complete streets doesn't address transit." The respondent went on to argue that, "For age-friendly communities, transit is essential. This is a failing of the policy" because of the "silos of information" that often disconnect transit planning from roadway design and funding. Respondents felt that a more explicit focus on transit integration in the policy with the new street design framework was necessary to improve transit opportunities for older adults.

POLICY ADOPTION: THE IMPACT OF COALITION-BUILDING AND PARTNERSHIP

While the complete streets philosophy in Louisiana at present does not address all components of age-friendly communities, the potential for positively improving the road connections of local communities provided the foundation for the coalition of advocates to push for change within the DOTD. As mentioned in the timeline of key events above, the Louisiana complete streets policy at the state level was passed as an administrative action by DOTD in 2010. This action was preceded by a concerted process of engagement and coordination through a legislatively enabled working group that brought key stakeholders together to determine the best path forward. With a diverse group working to define feasible goals, this process created both momentum and a defined platform for change.

This process of official and diverse stakeholder engagement in the drafting of DOTD policy was uncommon and opened unique opportunities for significant change within the institution. Interviews with the stakeholders revealed two key themes that emerged from the policy adoption process: the openness of DOTD leadership in the policy adoption process and the importance of AARP leadership for encouraging age-friendly policy change.

DOTD Leadership Open to Change

One of the most difficult processes an organization can go through is refining the core mission of the organization. Members of the organization establish routine practices and procedures that govern how the organization acts in relation to the broader organizational goals. When the underlying organizational mission is altered, the routine practices and culture of the organization must be changed to meet these new goals (Denhardt, Denhardt, & Aristigueta, 2009; Latham, 2003).

Most successful culture change exercises have the buy-in of the internal management of the organization. The goal of culture change emerges from management discussions and is then "applied" through internal leadership channels (Denhardt et al., 2009). The Louisiana case fits this model because higher-level leadership was supportive of the complete streets policy changes. Respondents consistently pointed to the buy-in of leadership members, as one MPO respondent put it, "at the top level" pushing for change.

The result is, as one of the non-profit respondents argued, that DOTD acted as a "partner" and "walked with us through this journey" of complete streets policy adoption. This coordination in policy adoption is rare and an important model for future policy adoption efforts.

AARP Leadership Created a Credible Partner

While the DOTD leadership was open to change, they were strongly encouraged to conduct an engagement process that brought new voices to the transportation policy discussion. One of the leading voices for change in this engagement process was AARP, a powerful advocacy organization for the needs of older Americans.

While a number of local organizations had worked towards changing the LADOTD and local DPW practices for years, the addition of AARP in active transportation policy discussions added significant political clout. AARP set the framework for policy change in motion through its Livable Communities Council. The Council brought together a set of Louisiana stakeholders to search for key actions to improve Louisiana communities and eventually focused on pushing for a complete streets policy as a key step in improving transportation choices for older Louisiana residents. AARP took this policy aim and helped to create a strong legislative agenda to push for a study resolution to examine how complete streets policy might work in Louisiana. Instead of a more narrow focus on alternative transportation, AARP, as one of the consultants argued, did "a great job reframing how we talk about this stuff" to include age-related and public health issues.

This broader approach and the push from AARP provided a credible and powerful voice for change that active transportation advocates alone could not provide. One of the MPO respondents argued that:

> The foundation was strong with having a large influential group like AARP pushing this . . . They used political status to get this passed (and) kept DOTD motivated . . . The great lesson is having the support and weight of an organization like AARP, which allowed us to have the legislative mandate for DOTD to be doing this.

POLICY IMPLEMENTATION: TURNING AN "OCEAN LINER"

Despite impressive results in terms of policy adoption, the policy implementation phase has been slow. According to a DOTD respondent, "The culture itself takes a while to start changing" and is like changing the course of an "ocean liner." Turning a massive organization in a fundamentally different direction requires changing the daily practices and regulations that govern hundreds of employees across the state. These changes are often slower than advocates for policy change would like.

Transforming the policy template into concrete results in projects has proven to be the central administrative challenge of the Louisiana complete streets policy. Respondents repeatedly pointed to the culture of the DOTD as one of the barriers to implementation but noted that this culture appeared to be changing. These interlinked themes of slow implementation and a resistant but changing DOTD culture are analyzed below.

Slow Implementation

While the administrative policy was passed in 2010, implementation of the policy in the three years since passage was broadly seen as "very slow" (nonprofit respondent) and "ad hoc" (transportation planner) across the state. The result has been that some parts of the state have seen strong implementation and change, while others have seen little in the way of measurable progress.

New Orleans, for example, has been able to pass a local complete streets ordinance and create a robust institutionalization process for complete streets. Much of the focus has been centered on rebuilding damaged infrastructure from Hurricane Katrina through the federal submerged roads program. While the disaster-focused federal funds were limited to rebuilding infrastructure to pre-Katrina conditions, the city was able to use local funds and an infusion of $10 million in American Recovery and Reinvestment Act funding to enhance roadways to meet new complete streets standards (Lee & Hall, 2011). This process propelled New Orleans forward as a leader in complete streets implementation in Louisiana with a targeted use of continental crosswalks and countdown clocks to improve pedestrian crossings and 78 miles of new bicycle facilities installed since Hurricane Katrina.

In much of the rest of the state, a much more ad hoc approach to implementation has been the norm. This ad hoc and incomplete policy implementation was seen as resulting from a lack of a "strategic approach" (transportation planner) to implementation without clear rules and guidance. There has been a lack of "measurable outcomes" (MPO respondent) and "goals for the program" (public health respondent), resulting in a "policy implemented differently in different places" (consultant respondent).

One MPO respondent speculated that the differential implementation resulted from different levels of political support within government organizations and "more developed advocacy base(s)" in some locations. For example, the respondent pointed out that key institutional actors across the state like "DOTD district staff, MPO, municipal staff, and citizen/advocates" had differing levels of knowledge of complete streets policies. These differences in knowledge defined "what type of implementation impact you'll find in different parts of the state." The political support of key institutions, knowledge of the policy among actors, and advocacy base pushing for change appear to be key components that define the localized differences in how the policy is being carried out in communities around Louisiana. This same institutional framework for complete streets change is likely necessary to propel complete streets forward in other areas. More research is needed to establish this as a consistent framework across states.

DOTD Culture

While multiple institutional actors define the political calculus of complete streets policy implementation in Louisiana, the key organization identified in the interviews was the DOTD itself. There was broad consensus among respondents that administrative changes were necessary for the DOTD to meet the needs of older citizens. Underlying the technical changes necessary to create more age-friendly communities, respondents pointed to the need for a broad culture change in the organization's basic practices and perceived mission.

Respondents felt that the DOTD was not yet focused on creating age-friendly communities, with one consultant respondent arguing that age-friendly communities were not yet on DOTD's "radar." The DOTD, many respondents felt, basically still sees its mission as the provision of highways. One urban planning respondent described DOTD's basic orientation as a "culture of moving cars."

These perceptions were broadly held by respondents, with nearly all interviews focusing on the need for the DOTD to broaden its perceived mission and orientation towards change. A set of respondent answers helps to highlight this. For example, one of the public health respondents argued that the DOTD uses a "very narrow approach" and "could stand to be more flexible." One of the planning respondents argued that division staff located away from headquarters appeared to "have trouble with change" and to represent "old school" thinking. Finally, an MPO respondent noted that leadership support was evident, but "I don't think that level of support is translating down to every level that it should, resulting in people not understanding what they need to do to be part of a new paradigm." The MPO also noted that complete streets was still seen by many "as an add-on, not an integral part of the project."

All of this appears to translate into what one consultant respondent referred to as a culture "in transition," with differing levels of the organization moving towards complete streets policy implementation and other sectors resisting change. One of the nonprofit respondents noted that the policy change was only the first step in changing the culture and practices of the organization. The respondent noted that "Culture change doesn't happen by itself." Instead, the respondent argued, "You need people to do that. You need a spirit of culture change and an understanding within the DOT about how culture change takes place." The respondent maintained that "tension" is sometimes necessary to create changes within an organization. The nonprofit respondent argued that, "Tension isn't necessarily bad. If you are comfortable with what you do every day for 30 years, why would you be interested in complete streets? What is it that motivates you to get going on this?"

The answer from respondents to these questions was that leadership from the top of the organization needed to provide clear direction about the precise meaning and intention of the policy. One of the non-profit respondents argued that it was "necessary" to work on "training of the engineers and practitioners within the state agency . . . to nurture that culture." One public health respondent summarized the key implementation steps as the "need to train, document progress, and work with communities to explain what they can expect."

Most respondents felt that these steps could help to improve the transition to a fully functioning complete streets policy and transform ad hoc practices into what one of the consultant respondents referred to as "standard operating procedure."

DISCUSSION

The complete streets philosophy of providing improved access to destinations for people of all ages and abilities appears to offer an important opportunity for age-friendly advocates to begin changing the transportation choices of older residents. Interview data from Louisiana reveal that the complete streets approach has gained traction as a compelling and potentially powerful tool for helping to build communities that will allow older residents to age in place. The Louisiana complete streets policy adoption process was both highly ranked and innovative, with multiple voices that had previously been excluded from the transportation policy debate included directly in the policy adoption process. The presence of AARP as the lead coalition member appeared to expand the political base of support and provide an opportunity for the DOTD to begin to change practices.

While the policy adoption process appeared successful, the implementation of the policy has been frustratingly slow for many age-friendly advocates. The culture of the main agency tasked with the provision of transportation in Louisiana, the state DOTD, has for years been dominated by auto-service delivery interests, leaving pedestrians, bicyclists, and transit-users without adequate accommodations. Changing these entrenched practices has proven difficult with differences in policy changes reported around the state.

One of the primary reasons for differential policy implementation across the state is that there are localized differences in the knowledge of complete streets by the DOTD districts, MPO and city staffs, and local advocates. This has left the single-state written policy to be interpreted through all of these local power dynamics, resulting in multiple policy interpretations around the state. Respondents were clear in arguing for a much more systematic approach to implementation that brought the policy out of the offices of DOTD leadership in Baton Rouge and down to the streets in local Louisiana communities. Respondents felt that this more systematic approach should include clear guidance from top leadership on implementation expectations, improved training, documented results of the process, and improved engagement with the community to outline expectations for the policy change.

From an institutional perspective, analysis of the complete streets policy process in Louisiana suggests that overcoming bureaucratic inertia through coalition building and engagement is a key step in building more age-friendly communities. This process involves numerous organizational changes at the DOTD. Leadership focus on these fundamental issues could help manage pockets of resistance to the policy and help create a less ad hoc policy culture within the organization (Millar & Sundquist, 2012).

While complete streets policy implementation is still ongoing in Louisiana, interview respondents were encouraged by the changes taking place at the DOTD. One of the consultants interviewed for this study summarized the feelings of nearly all of the respondents by noting that, "There are a lot of people at DOTD who would like to see this happen . . . The culture of DOTD is changing and in the future it won't give as much of a problem as it is now." These changes present opportunities to begin to re-envision how to build age-friendly communities in Louisiana.

More broadly, the Louisiana case suggests that creating more age-friendly communities will involve creating innovative platforms for nongovernmental organizations to engage with the governmental organizations tasked with providing transportation services. At present, these transportation organizations often do not see their core mission as helping to build age-friendly communities. This case suggests that engagement, especially through a powerful advocacy organization like AARP, offers an important avenue that can help to build the types of ongoing partnerships necessary to improve age-friendly community transportation choices.

REFERENCES

Bailey, L. (2004). *Aging Americans: Stranded without options*. Washington, DC: Surface Transportation Policy Project.

Bassett D.R., Jr., Pucher, J., Buehler, R., Thompson, D.L., & Crouter, S.E. (2008). Walking, cycling, and obesity rates in Europe, North America, and Australia. *Journal of physical activity & health*, *5*(6), 795–814.

Brown, R.R. (2008). Local institutional development and organizational change for advancing sustainable urban water futures. *Environmental Management, 41*, 221–233.

Burgess, S., Purvis, A., Moore, A., Chandra, L., Netherway, C., & Richards, P. (2010). *Complete streets guidelines for urban street design*. Fortitude Valley, QLD: Institute of Public Works Engineering, Australia Queensland Division.

Complete Streets for Canada. (2013). *Complete streets comes to Edmonton*. Last accessed August 27, 2013. Available from http://completestreetsforcanada.ca/news/complete-streets-comes-edmonton.

Denhardt, R.B., Denhardt, J.V., & Aristigueta, M.P. (2009). *Managing human behavior in public and nonprofit organizations*. Los Angeles: Sage.

Downey, M.L., III. (2007). *Legislative Considerations for Long Term Policy Change. In National Roundtable on Surface Transportation*. New York: Regional Plan Association and Lincoln Institute for Land Policy.

Ernst, M. (2011). *Dangerous by design 2011: Solving the epidemic of preventable pedestrian deaths*. Washington, DC: Transportation for America.

Farber, N.E. (2011). *Aging in place: A state survey of livability, policies and practices*. Washington, DC: AARP.

Fields, B., & Hull, T. (2013). Policy implications of the nonmotorized transportation pilot program. In John Renne and Billy Fields (Eds.), *Transport Beyond Oil* (pp. 265–290). Washington, DC: Island Press.

Griffin, G. (2011). Evolving innovations in bicycle transportation. *Practicing Planner* (9), 3.

Handy, S.L., McCann, B., Bailey, L., Ernst, M., McRee, L., Meharg, E., . . . & Wright, K. (2009). *The regional response to federal funding for bicycle and pedestrian projects*. Davis, CA: Institute of Transportation Studies, University of California, Davis.

He, W., Sengupta, M., Velkoff, V.A., & DeBarros, K.A. (2005). *65+ in the United States: 2005*. Washington DC: U.S. Census Bureau.

Healey, P. (2004). Creativity and urban governance. *Policy Studies, 25*(2), 87–102.

Johnson, B.J., & White, S.S. (2010). Promoting sustainability through transportation infrastructure? Innovation and inertia in the Kansas City Metropolitan Area. *Journal of Urban Planning and Development*, *136*(4), 303–313.

Kerr, J., Rosenberg, D., & Frank, L. (2012). The role of the built environment in healthy aging community design, physical activity, and health among older adults. *Journal of Planning Literature*, *27*(1), 43–60.

Latham, G.P. (2003). Goal setting: A five-step approach to behavior change. *Organizational Dynamics*, *32*(3), 309–318.

Lee, L.J., & Hall, B. (2011). Louisiana's recovery. *Public Roads*, *75*(1), 5.

Lehning, A.J. (2012). City governments and aging in place: Community design, transportation and housing innovation adoption. *The Gerontologist*, *52*(3), 345–356.

Lynott, J., Haase, J., Nelson, K., Taylor, A., Twaddell, H., Ulmer, J., . . . & Stollof, E.R. (2009). *Planning complete streets for an aging America* (No. 2009–02). Washington, DC: AARP Public Policy Institute.

Mapes, J. (2009). *Pedaling revolution: How cyclists are changing American cities*. Corvallis, OR: Oregon State University Press.

Miles, M.B., & Huberman, A. M. (1994). *Qualitative data analysis: An expanded sourcebook*. Thousand Oaks, CA: Sage Publications.

Millar, R., & Sundquist, E. (2012). *The innovative DOT: A handbook of policy and practice*. Smart Growth America and the State Smart Transportation Initiative.

Pucher, J., & Dijkstra, L. (2003). Promoting safe walking and cycling to improve public health: Lessons from the Netherlands and Germany. *American Journal of Public Health*, *93*(9), 1509–1516.

Seskin, S., & Gordon-Koven, L. (2013). *The best complete streets policies of 2012*. Washington, DC: National Complete Streets Coalition and Smart Growth America.

Snyder, T. (2013). *500+ complete streets policies in place, but not the most important one*. Streets blog August 16, 2013. Available from http://dc.streetsblog.org/2013/08/16/500-complete-streets-policies-in-place-but-not-the-most-important-one/

Transportation for America. (2011). *Dangerous By design: Louisiana solving the epidemic of preventable pedestrian deaths (and making great neighborhoods)*. Available August 12, 2013, from http://t4america.org/docs/dbd2011/la-dangerous-by-design.pdf

Wahl, H.W., Iwarsson, S., & Oswald, F. (2012). Aging well and the environment: Toward an integrative model and research agenda for the future. *The Gerontologist*, *52*(3), 306–316.

Wells, C.W. (2013). *Car country: Automobiles, roads and the shaping of the modern American landscape, 1890–1929*. Madison, WI: University of Wisconsin.

15 Age-Friendly Cities and the WHO Checklist

Lessons from a Portuguese Survey

Alexandra Lopes, Teresa Pinto, and Rute Lemos

INTRODUCTION

This paper will bring to light some of the findings from the *cIDADES* survey performed in Portugal between 2010 and 2011. The name of the research project, *cIDADES*, represents the coming together of two Portuguese words: the word for "cities," *Cidades*, and the word for "age groups," *Idades*. The aim of the survey was to assess the awareness of senior citizens of how many of the age-friendly features identified on the Checklist of Essential Features of Age-Friendly Cities developed by the World Health Organization (WHO) were actually available in their cities. The findings of the survey are discussed from a methodological standpoint and from a more substantive perspective.

The objective of this paper is two-fold: (1) to analyze the performance of the checklist when used to measure how Portuguese senior citizens assess their cities, with particular emphasis on the dangers of oversimplification associated with the use of the checklist as currently organized into eight dimensions, as well as bringing up the policy design implications of the multidimensionality of each domain; (2) to draw on the results of the survey to discuss trends in how aging issues have intersected with urban planning in Portugal and demonstrate the coexistence of two different, though complementary, trends. One places greater value on policies that target the physical environment but less on age-group-targeted approaches, and the other suggests a spiral effect in planning for age-friendliness that translates into integration and comprehensiveness in the development of age-friendly environments. These two strands are proposed as criteria for consideration when using the checklist.

POLICIES OF PLANNING FOR ALL AGES

Urban planning and administration are defined by a set of core goals that ultimately aim at improving the welfare of people and communities. This is achievable if the environments where people live are experienced as convenient, equitable, healthy, efficient, adequate, attractive, and meaningful, or, in other words, when the planned environments meet people's needs, expectations, and preferences.

An aging population, together with a growing awareness of the need to bring the framework of active aging into every field of public policy, has led to the inclusion of age, and old age in particular, into the concept of inclusive urban planning (Burton & Mitchell, 2006). This, in turn, has influenced the development of a systematic approach to healthy and active aging in urban environments, which led to the 2006 global movement for age-friendly cities promoted by the WHO. The resulting publication, the *Global Age-Friendly Cities: A Guide*, and its companion, the *Checklist of Essential Features of Age-Friendly Cities* (hereafter named "checklist"), offer a set of guidelines to engage and assist cities in becoming more "age-friendly." Among other goals, the checklist was designed to provide a tool for urban planners to assess the age-friendliness of urban environments and to be applied to planning for all ages, thus eliminating age-related obstacles that prevent inclusiveness, participation, healthy aging, and the quality of life of senior citizens.

City planners and city managers have generally been open to the inclusion of age-friendliness as a guiding principle of policy design. Many cities around the world have joined the movement and have developed a variety of initiatives to create age-friendly environments and foster the inclusion and participation of senior citizens in urban issues. The checklist in particular has been widely used as a reference framework for communities to engage in the self-assessment of how age-friendly they are (Green, 2012).

One of the core principles of the age-friendly movement emphasizes the need to involve everyone in all of phases of policymaking—from needs analysis to the development of solutions, as well as including the implementation and monitoring/assessment of results. The checklist itself emerged from a methodology that involved the consultation of stakeholders, mostly by conducting focus groups with different social groups, ranging from senior citizens to professionals and volunteers. In many cities that have joined the age-friendly movement, the use of the checklist has remained very faithful to the methodological orientation of its initial formulation, which draws on a qualitative participatory research design. In several of these cities, this has evolved into the use of focus groups and qualitative interviewing of small samples of individuals and stakeholders to discuss needs and priorities, to debate policy alternatives, and to discuss the results of specific initiatives.

Starting from the policy recommendations set out in the checklist, this research project (*cIDADES*) was launched as an action-research initiative in Portugal in 2010.[1] The main goal of the project was to assess the age-friendliness of urban environments in Portugal and bring together the points of view of various stakeholders, most importantly, the points of view of senior citizens themselves. However, rather than debating the ideal appearance of age-friendly cities and drawing on the proposals by authors such as Buffel and Phillipson, our theoretical starting point involved moving slightly away from the dominant paradigm of identifying the features of an ideal urban environment and instead focused on assessing the existing characteristics of cities and, moreover, on how individuals experience the spaces where they live. This is a theoretical standpoint that emerges from a critical analysis of the age-friendly movement. Critics argue that "age-friendly cities" is a consensus-based concept, which runs the risk of masking the existing tensions and contradictions of urban planning in relation to the age-friendly debate (Buffel, Phillipson, & Sharf, 2012). In this sense, more than identifying what people would consider adequate, this point of view highlights the importance of assessing how individuals experience and perceive the environments in which they live and the factors that determine variations in those experiences.

From a methodological standpoint, this involved a comprehensive approach to living environments and implied consulting with a broad range of people to gather their perceptions of the degree to which the places in which they live are age-friendly. In the ensuing sections, findings from a survey performed in Portugal to assess the degree of age-friendliness of cities from their inhabitants' perspectives are discussed. The starting assumption is that individuals interact with the space where they live and they forge their life experiences within that interaction. These experiences define life trajectories, create expectations, and give meaning to situations. Therefore, more than measuring the existing features of space, the underlying approach carried out in *cIDADES* involved identifying what individuals see as being the features of the space where they unfold their life narratives.

AGING AND URBANIZATION IN PORTUGAL

The interest in urban questions related to aging made its first appearance in Portugal soon after the age-friendly city concept appeared on the international stage. As in many countries in Europe and worldwide, Portugal has been experiencing a rapid demographic change, with older cohorts growing at a very fast rate. According to the results of the 2011 census, the share of those aged 65 or older was 19.3% of the total population, which represents an 18% increase compared to the same age group in 2001 (INE, 2012). Portugal ranks high on the list of the most aged countries in the world, according to the estimates of the United Nations. In 2012, it was ranked as the seventh lowest country in terms of its old-age support ratio, with a little over three working-age adults for each older person[2] (UN, 2012).

The increasing weight of older cohorts, coupled with the growing public visibility of their specific needs, especially those not met, has brought the topic of aging into all the fields of public policy, including urban planning and management. A heightened awareness that aging may constitute an obstacle to social inclusion has quickly taken center stage in the political agenda and has opened the way for a diversity of initiatives across both civil society and institutional agents. The creation of the Global Network of Age-Friendly Cities by the WHO has also echoed in Portugal and has attracted vivid interest among city managers. At the national level, the movement towards elder-friendly community development has brought the implementation of a National Network of Age-Friendly Cities managed by an NGO, *Associação VIDA*, which already works with 103 out of a total of 308 municipalities across the country. The goal of the network is to make Portugal the most age-friendly country in the world by encouraging the implementation of the policy recommendations of the WHO checklist. *cIDADES* is a broad action-research project that is part of the effort to put aging on the agenda of urban planning and urban management.

In this paper, the concept of "city" is used arbitrarily. Until recently, there was no commonly accepted definition of what constitutes a "city," which not only led to problems of international comparability and validity in cross-country studies, but also raised issues within countries when analyzing diverse territories. The recent OECD definition identifies cities as those with urban centers of at least 50,000 inhabitants, which for Portugal would mean 16 cities (Dijkstra & Poelman, 2012). Nevertheless, the Portuguese official definition of a city by the national authorities draws on mixed criteria and is not exclusively based on population size. Legislation dictates a series of official requisites for being considered a city, combining population (a minimum of 8,000 voters in a continuous population agglomerate, understood as an urban city center) with functional criteria (the community must have at least half of the facilities listed, including health care, culture, leisure, and education infrastructures, as well as public transportation and outdoor public spaces).[3]

For this study, the official Portuguese definition of a city was used, which means that relatively small communities, both in terms of geography and population size, along with big urban centers like Lisbon or Porto, were included in the sample. Naturally, this raises issues about the reliability of the measurement when applied to such a diverse array of environments. Nonetheless, the discussion of such a contentious topic is not the aim of this paper. Instead, it is argued that the main goal of the *cIDADES* survey was to assess how people perceive the age-friendliness of the communities where they live, irrespective of the specific dynamics that take place in environments with different sizes, with different population densities, and with different levels of urbanization.

Moreover, Portugal's territory and urbanization have very specific characteristics. Due to historical and sociopolitical dynamics during the first three quarters of the twentieth century, Portugal experienced a very low rate of urbanization alongside an equally low rate of industrialization. The urban network is centered around two main urban areas (Lisbon and Porto) and highly concentrated along the coastline, with very clear inequalities in spatial distribution. Historically, there have been very few medium-sized urban centers, and the concentration of the population around Lisbon and Porto has continued to increase, which has given rise to a series of satellite towns surrounding the two metropolitan areas. Another historical trait characterizing the distribution of Portugal's population is the sharp divide between the sparsely populated North and the much more concentrated distribution of the population in the South (Gaspar, 1987). Both the design and the findings of the *cIDADES* survey take these characteristics into account.

GENERAL CONSIDERATIONS ON METHOD

The methodological approach involved in the research is described in detail in the appendix (Appendix to Chapter 15). Issues addressed include sampling, data collection procedures, and data analysis protocols.

cIDADES was launched in 2010 and involved a complex, mixed-method research design. The goal was to collect input from an array of social actors, ranging from academic experts and policymakers to ordinary citizens, in order to understand how different agents and stakeholders experience urban environments and how they assess the age-friendliness of the cities where they live and work. One of the phases involved a survey of a national sample of individuals aged 55 and older designed to understand how older populations experience their cities and to examine the current state of urban environments from the perspectives of their residents. The questionnaire used in the survey was based on the WHO checklist and was organized into eight sections, mirroring the dimensions of the global guide.[4]

The survey did not involve a pure, randomized sampling design since there was no access to population databases. The first stage in the sample design resorted to a stratified random selection of municipalities to provide a representative sample of the territories included. We worked with municipality-level units selected following the stratification of the national territory according to NUTs II (administrative units defined by law that divide the territory of the country in seven regions: North, Center, Lisbon and Tagus Valley, Alentejo, Algarve, Madeira, and Azores), population density, and aging rate. The strata for population density and aging rate were defined according to the official INE criteria (National Statistics Office). The final sample included data from 81 municipalities across the country and proportionally covered the territories as stratified. This number represents roughly 26% of the total municipalities in the country. The distribution of sampled municipalities by stratification criteria is proportional to the weight of each stratum in the population.

Based on the overall structure of the population of each participating municipality, a set of combined quotas were defined to account for: gender distribution, weight of those 55 plus in the total population of the municipality, and age groups. Each municipality received its quotas with the instructions to use them as references for the final composition of the local samples. In many cases, the quotas were not strictly followed, and the size of the final sample largely exceeded that requested. Data were collected with purposive sampling in the urban areas of each of the sampled municipalities. Questionnaires were administered in face-to-face interviews.

The final sample comprises 8,658 individuals, with local samples ranging from 50 to 1,117. No weighting procedure was used given the non-randomized sampling design. Furthermore, the data will only be used to test some assumptions about the performance of the checklist and to identify broad-ranging trends in urban planning in Portugal. We shall not be establishing any comparisons between cities, not only because the data do not offer valid grounds for such. but also because the protocol included a clause of anonymity for municipalities in the discussion of results.

MULTIDIMENSIONALITY IN THE ASSESSMENT OF AGE-FRIENDLINESS

The starting point for this study was the list of features that the WHO determined to be the features of age-friendly cities. These features are grouped into eight domains of urban life: outdoor spaces and buildings, transportation, housing, social participation, respect and social inclusion, civic participation and employment, communication and information, and community support and health services. Each dimension, in turn, includes several items that refer to specific characteristics of the community environment, both material and immaterial, that should be displayed by a city to be considered age-friendly. The questionnaire used in the Portuguese survey took the items listed under each of the eight dimensions as its basis. Each item was built into a sentence, and individuals were asked to state whether the content described in the sentence applied (or not) to the communities where they lived.

The fact that the checklist includes eight dimensions of age-friendliness already points to the multidimensional nature of urban environments. What the results of the survey performed in Portugal suggest, however, is that this multidimensionality may be a bit more complex than what is implicitly suggested by the original groupings in the checklist or, at least, interpreted from different points of view.

The principal components analysis performed on the items as formulated in the survey's questionnaire, and taking the original dimensions of the checklist as a starting point, have resulted in the identification of the following groupings of variables as listed in Table 15.1.[5]

The numerical solution suggests that, within each of the eight dimensions included in the checklist, it is possible to identify more specific groupings of items. Furthermore, the fact that these are uncorrelated dimensions within

Table 15.1 Dimensions of Grouping of Indicators Drawn from the Portuguese Survey and the Corresponding Dimension of Analysis in the WHO Checklist

Dimensions identified (Cronbach's Alpha)	Original groupings in the WHO checklist (KMO[a])
1. Outside space in general and green spaces (0.812)	Outdoor Spaces and Buildings (0.933)
2. Sidewalks (0.764)	1.
3. Crossing streets (0.636)	1.
4. Public buildings location and accessibility (0.647)	1.
5. Public buildings physical characteristics (0.819)	1.
6. Toilet facilities in public places (0.845)	1.
7. Public transportation frequency and coverage (0.787)	Transportation (0.891)
8. Public transports costs (0.609)	1.
9. Transportation for people with reduced mobility (0.666)	1.
10. Traffic and roads (0.405)	1.
11. Bus drivers attitude (0.541)	1.
12. Bus stops location and characteristics (0.885)	1.
13. Housing comfort and characteristics (0.796)	Housing (0.823)
14. Housing and adaptation to special needs (0.668)	1.
15. Leisure and events information (0.753)	Social Participation (0.804)
16. Leisure and events venues (0.676)	
17. Consultation of older people (0.853)	Respect and Social Inclusion (0.873)
18. Intergenerational values (0.847)	
19. Volunteering (0.653)	Civic participation and employment (0.827)
20. Age discrimination at work (0.913)	
21. Incentives to take paid work (0.828)	1.
22. Access to information in the media (0.773)	Communication and Information (0.931)
23. Age-sensitiveness of contents broadcasted by the media (0.866)	

24. Format of media contents (0.924)	1.
25. Information lines (0.836)	1.
26. IT availability (0.833)	1.
27. Health and social care services availability (0.831)	Community support and health services (0.896)
28. Access to health and social care services (0.869)	
29. Location of services (0.773)	1.
30. Service delivery and staff (0.847)	1.

[a] Keiser-Meyer-Olkin statistics of sample adequacy measure the total amount of variation in the data that can be explained as common variation.

each of the original WHO groupings suggests that the grouping of factors under an eight-category typology may be oversimplifying the dimensions of assessment of age-friendliness if one intends to use the checklist to look at the broad groupings. It is known that the WHO checklist includes a series of sub-dimensions within each of the eight factors, and it is acknowledged that these sub-dimensions are generally mirrored in the groupings found after analyzing the data from the Portuguese survey. However, to be considered sub-dimensions of the same overall topic, the sub-dimensions should still correlate strongly among themselves, which is not the case for Portugal. This fact may have strong implications regarding the use of the checklist for self-assessment of age-friendliness in general. If one takes into account the transferability impact within each of the eight dimensions when a city scores low and high on different aspects within one dimension, this is compounded. The data for Portugal, as displayed below, has shed some light on this risk.[6] Figure 15.1 displays the average scores computed for each of the dimensions identified in Table 15.1, as well as the range of variation between the first and third quartiles.

Average scores can be interpreted as a measure of achievement within each dimension, ranging from a minimum of zero, if all of the respondents state that all of the items included in the dimension are absent, to a maximum of 100, if all of the respondents state that all of the items included in the dimension are available in their cities. A score of 50% would then mean that, on average, individuals consider that only half of the items included in the dimension are available in their cities, which could be interpreted as a signal of moderate achievement in that specific dimension.

Let us look, for example, at the dimension related to outdoor spaces and buildings. The results shown in Figure 15.1 point to a moderate to high level of satisfaction with the conditions of use of outdoor and green spaces (52.6%), but a lower level of satisfaction with circulation within

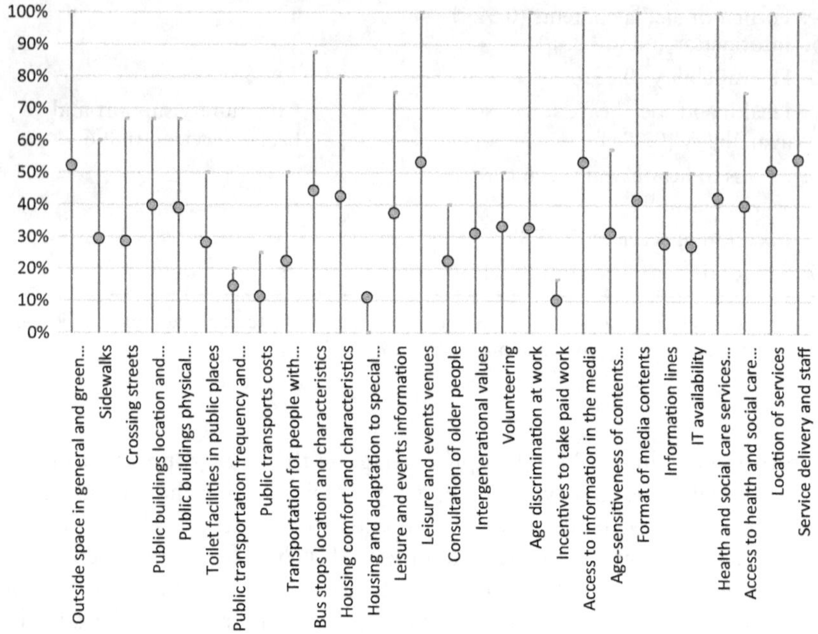

Figure 15.1 Average Score and Interquartile Variation Range for Each Dimension of Analysis in Age-Friendliness

them (29.4% is the average score for the characteristics of sidewalks, and 28.6% is the score for pedestrian crossing areas). Transportation is another example to look at. Public transport support facilities (location and material characteristics of bus stops) get a relatively high level of satisfaction (44.4%). However, satisfaction ratings are much lower for the frequency/coverage, cost, and accessibility of public transport for individuals with reduced mobility (14.5%, 11.4%, and 22.5%, respectively). If grouped under the same rubric, these indicators would have transferred scores among themselves and would probably have resulted in a moderate overall assessment of the degree of age-friendliness of public transport in Portuguese cities. Moreover, the data show that there are different aspects within the same broad dimension that can be thought of as specific domains for policy design in urban sites in Portugal, especially because they involve the use of different resources and the mobilization of different agents.

The argument in this respect, and the first finding of the survey, is that there is something to be gained from a more detailed analysis of each of the eight dimensions of the checklist if one is to use the checklist as a tool for assessing age-friendliness of urban environments and even more so if one hopes to draw implications for policy design and priorities.

This paper does not contend that the groupings found for the Portuguese case can be extended to other cities worldwide. Also, it does not contend that the groupings found are substantially different from what can be found in the detailed description of each dimension in the original checklist. On the contrary, it makes a case for the need to examine how the tool can be adapted and used to provide a more accurate measurement for assessing the age-friendliness of cities in each national context. Taking the checklist as developed by the WHO as a starting point, it is possible to look for alternative criteria to work with the groupings found in the data. The broad eight dimensions found in the WHO checklist seem to draw on the theoretical aggregation of the different topics according to their broad field of action. In this light, all the topics listed under "Transportation" do in fact make a good deal of sense, since they are all theoretically related to transportation. The approach in this research involved looking for a different rationale when analyzing the groupings of items identified as statistically significant in the data. It involved a rationale that could assist the discussion of meanings of the groupings in terms of the intricate policymaking process.

Each dimension displayed in Figure 15.1 was classified according to its content (material or social nature) when considering the type of policy required if one was to target a specific dimension (e.g., outdoor spaces and green areas refer to a material domain of intervention, but bus drivers' attitudes point to an immaterial intervention, in this case, training and education). Each dimension was further classified according to the scope of its impacts (more general/undifferentiated or more targeted/group-specific nature), taking into account the centrality of the old age group (e.g., outdoor spaces and green areas refers to the general population, but adaptation of housing to people with special needs points to a more relevant impact among the older cohorts of the population). Table 15.2 displays the classification used by the authors to label each dimension under the two-dimensional categories. The choices of the authors are debatable in some cases, but the exercise was done to simplify the discussion and to draw conclusions from the data in terms of policy implications and needs.

Table 15.2 Classification of Each Dimension of Analysis of Age-Friendliness According to the Scope of the Intervention Required in Policy Design (Material/Immaterial) and According to the Target Population Relative to Older People (General/Specific)

	Scope of intervention		Target-population	
	Material	Immaterial	General	Specific
Outside space in general and green spaces	X		X	
Sidewalks	X		X	
Crossing streets	X			X
Public buildings location and accessibility	X		X	

(*Continued*)

Table 15.2 (Continued)

	Scope of intervention		Target-population	
	Material	Immaterial	General	Specific
Public buildings physical characteristics	X		X	
Toilet facilities in public places	X			X
Public transportation frequency and coverage		X	X	
Public transport costs		X	X	
Transportation for people with reduced mobility	X			X
Traffic and roads	X		X	
Bus driver attitudes		X	X	
Bus stop locations and characteristics	X		X	
Housing comfort and characteristics	X		X	
Housing and adaptation to special needs	X			X
Leisure and events information		X	X	
Leisure and event venues	X		X	
Consultation of older people		X		X
Intergenerational values		X		X
Volunteering		X		X
Age discrimination at work		X		X
Incentives to take paid work		X		X
Access to information in the media		X	X	
Age-sensitiveness of contents broadcasted by the media		X		X
Format of media contents		X		X
Information lines	X		X	
IT availability	X		X	
Health and social care services availability	X		X	
Access to health and social care services		X	X	
Location of services	X		X	
Service delivery and staff		X	X	

Following the classification displayed in Table 15.2, the mapping of the averages from Figure 15.1 was then processed. Averages were plotted against a two-dimensional space. The horizontal axis in the figure represents the material/social dichotomy. The vertical axis represents the general/specific orientation in terms of population-target. Material intervention and general target population assignments are, therefore, plotted on the positive axes, while immaterial intervention and specific target population assignments fall on the negative axes. As an example, the 52.6% average score noted earlier for "outdoor spaces and green areas" is plotted in the upper right quadrant because it was classified as material intervention for the general population. On the other hand, the 22.3% average score noted earlier for "consultation of senior citizens" falls in the lower left quadrant, given that it has been classified as an immaterial intervention targeted to the specific group of senior citizens. The results are shown for all the dimensions in Figure 15.2.

Plotting the dimensions of assessment of age-friendliness in a two-dimensional space reveals trends that are worth discussing.

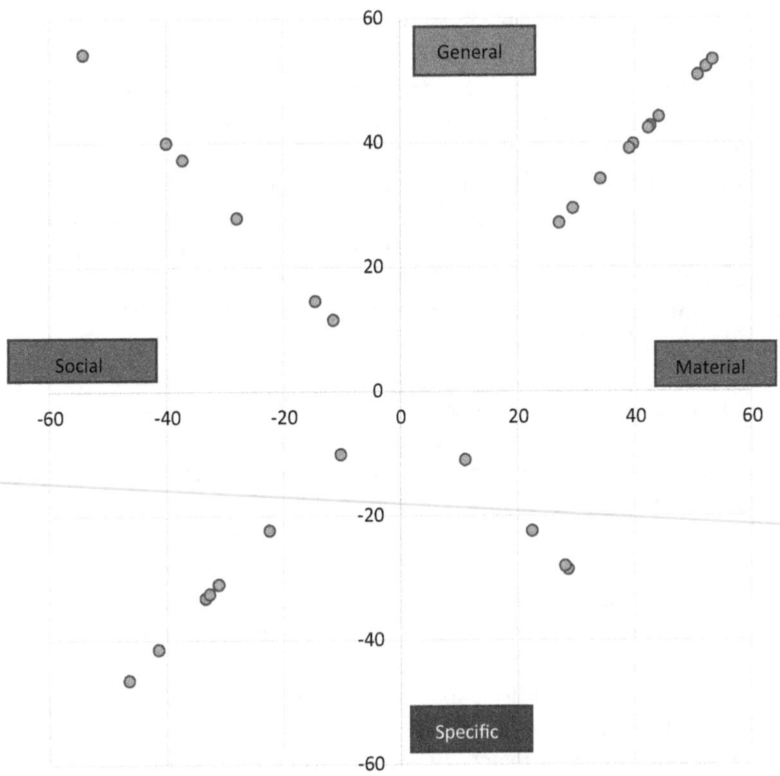

Figure 15.2 Dimensions of Age-Friendliness According to a Two-Dimensional Classification (Scope of Intervention and Target Population)

Overall, it can be seen that there is a higher level of satisfaction with those features that involve material aspects of the city and that involve policies of a more general and standardized nature. This would correspond to the positive quadrant of the plot on the top right-hand side. Generally speaking, all of the dimensions that fall in that space have relatively satisfactory average scores. The same statement does not hold when we look at the other three quadrants. Although there is some variation within each quadrant, overall, there is a lower level of satisfaction when issues at stake involve policies of a more immaterial nature and when policies are more group-oriented. It is not so much the specific values that are of interest for our analysis, but what this pattern means in terms of policy design, the topic of the next section.

GENERAL TRENDS IN PORTUGUESE CITIES TOWARDS THE CREATION OF AGE-FRIENDLY ENVIRONMENTS: PHYSICAL ENVIRONMENTS AND UNDIFFERENTIATED USERS

On taking a closer and more detailed look at the scores of each of the dimensions and combining them with the information displayed in Figure 15.2, the results suggest that there is room to consider a dominant orientation in how local managers and city planners have addressed the question of age-friendliness in Portugal. That trend points to a continuing orientation of the mainstream towards policies that address physical environments and broad age-disconnected policies.

It is generally accepted in the literature that the factors that contribute to a city's age-friendliness involve both physical (the built environment) and social factors (social and civic engagement) (Lui, Everingham, Warburton, Cuthill, & Bartlett, 2009). The balance of these factors is important for creating favorable environments in which to grow old (Verté, 2012).

On the other hand, there are frequent calls, in this same literature, for the need to think about policymaking as a process in which competing interests come together and in which stakeholders exert pressure towards achieving what they perceive to be their specific goals. According to Harvey (2008), one of the key challenges for contemporary urban planners is to juggle the competing interests of different stakeholders in environments that are increasingly dominated by consumerism, tourism, and cultural industries. This may, in fact, draw attention away from the specific needs of some in favor of others and distort the principle of meeting everybody's needs.

The Portuguese survey results, as displayed in Figures 15.1 and 15.2, highlight greater achievements in relation to the built environment (public buildings, gardens and outdoor spaces obtained 39% and 52%, respectively) and the need to provide good levels of services to the overall population (friendliness of staff, leisure activities and cultural events obtained 54% and 53%, respectively). These factors are of great value and may provide the grounds for critical questions, namely those about the pressures on urban environments and on urban planning. The results show higher levels of dissatisfaction precisely in those areas related to the specific needs of senior citizens (problems related to reduced mobility and to circulation difficulties in public space, 11% in housing adaptation, 22% in transportation for people with reduced mobility, and 28% in crossing streets), as well as to the empowerment and participation of that age group (22% for consultation of senior citizens and roughly 10% for participation in the labor market).

Age-friendly communities are those that actively involve, value, and support senior citizens, both active and frail. Moreover, age-friendly communities are those able to accommodate effectively the changing needs of aging individuals and foster an inclusive relationship with both the physical and the social and symbolic environments. In this respect, there is a clear need for some balance in how age-friendly principles are incorporated by urban planners and city managers in Portugal. There is, in particular, a great need to reconcile the mainstream pressures arising from the political economy of the city to accommodate the needs and expectations of senior citizens. The political discourse has been dominated by the importance of developing community infrastructure and design specifications for various aspects of the built environment, but not necessarily tackling the specific needs of the aging population and even less so of frail elders. Less attention has been paid to the "social quality" and social relations that promote inclusion and participation. The way individuals experience their communities, as measured by our survey and displayed in the previous figures, points to this precisely.

As Buffel and colleagues highlight, it is critical to link the concept of age-friendly communities to that of urban citizenship (Buffel et al., 2012). Citizenship in turn is very much related to participation and equal access, which can be achieved only if the tendency to more or less universal standardized approaches in policy design is reconciled with pluralism and differentiation.

INTEGRATION OF POLICIES AT THE LOCAL LEVEL: SPILLOVER EFFECTS IN URBAN POLICIES FOR AGE-FRIENDLINESS

When discussing these issues at the territorial level, one needs to take into account the differences that exist among territories. So far, the results that have been addressed relate to a national level. To detail the analysis a bit further, it is important to look at some of the dynamics that take place at the local level. The main trend that was identified and that shall be discussed in this section of the paper points to a degree of integration in the overall orientation to age-friendliness. In other words, the cities that show higher scores in the different dimensions tend to take the leading positions in all of the aspects. This suggests the existence of a spillover effect in planning for age-friendliness.

Considering the dimensions that have been analyzed so far (as shown in Figure 15.1), comparative analyses across the sampled municipalities were performed, and the results were then grouped according to the combined average scores in each dimension. This was done by resorting to a rather simple procedure that classifies them into a three-group typology according to their dominant trends (high, moderate, and low average scores in dimensions of analysis). A 3-means cluster analysis with iteration (for technical details on this, see the methodological appendix[7]) was the method used. The results appear in Figure 15.3 in a radar plot that displays the average score per dimension in each group. The groups vary in size; the highest scoring group included 21 municipalities, and the lowest included 32 municipalities (the intermediate group was the largest—89 municipalities). The lines are the result of linear segments joining the average scores per dimension in each group.

The scope of this research permits neither a detailed analysis nor the identification of the municipalities of each group. Nevertheless, the overall analysis of the combined means for each dimension across the three clusters shows that what distinguishes the clusters is a consistent difference in the value of the mean scores without any intersection of clusters. In other words, there is a group of municipalities that consistently takes the lead in all of the dimensions, a second cluster that receives more moderate scores also in all of the dimensions, and a third group that consistently scores below average across the board. The three lines evidence parallel star-shaped effects, which present the highs and lows that consistently show up across all of the municipalities. The absence of any crossover lines in Figure 15.3

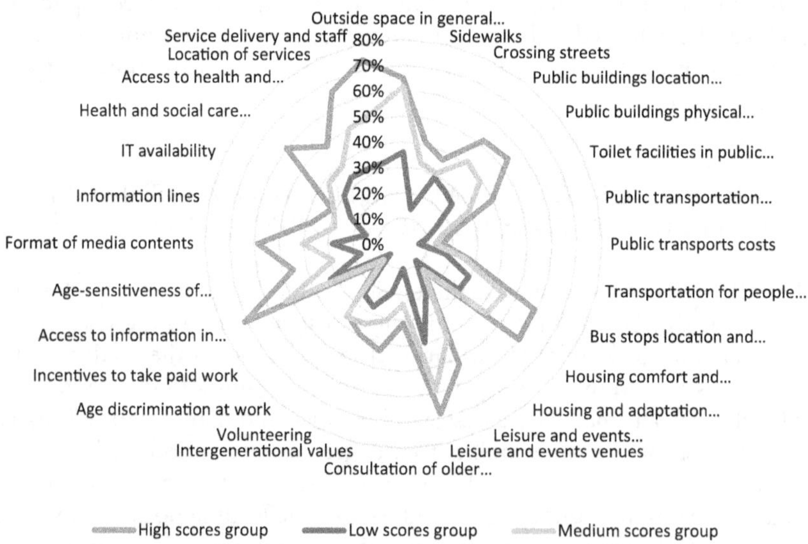

Figure 15.3 Radar for the Average Scores per Cluster of Sampled Municipalities

Table 15.3 Proportion of Sampled Municipalities within Each Scoring Group by Rural/Urban Classification and by Inland/Seaside Location (n)

Scoring cluster	Location of municipalities		Landscape is predominantly. . .	
	Inland	Along the costal line	%urban	%rural
High scores group	33.3 (7)	66.7 (14)	42.9 (9)	57.1 (12)
Low scores group	36.0 (32)	64.0 (57)	44.0 (39)	56.0 (50)
Medium scores group	34.3 (11)	65.6 (21)	43.7 (14)	56.3 (18)

shows that this spillover effect in policy design does not conflict with the previously discussed lack of integration of policies addressing age-friendly environments. It reinforces the belief that policymaking and urban planning evolve in a spiral fashion and that, once the process is in motion, it tends to spread to different dimensions of community life. This is an important finding since it reinforces the argument about the benefits of thinking about age-friendliness, irrespective of your starting point, and given the potential spillover impact across all of the areas of city planning.

The data were further inspected to assess if there was any variation in the orientation towards age-friendliness that can be explained by the characteristics of the territories. Differences were analyzed, controlling for location of cities (along the coast line vs. inland) and for type of landscape (predominantly urban vs. predominantly rural). Table 15.3 shows the results for the cross-tabulation.

From the analysis of the distributions within the two controlling classification criteria, the scoring groups are scattered across the country in what appears to be a rather random distribution, in both predominantly rural and urban settings, both inland and along the coastline. This is significant in itself, as it shows that issues of urban planning are probably not so dependent on the intrinsic characteristics of the territory but more on the willingness of decision-makers to implement age-sensitive policies.

CONCLUSIONS

Raising awareness of age issues in urban planning and city management is a clearly won battle, to our mind, especially in Portugal. International guidelines setting standards about what an age-friendly community should be are well established and have been duly incorporated into the political agenda.

The analysis of the Portuguese case leads us to believe that it is necessary to fine-tune the instruments used for assessing and monitoring how inclusive and overarching the age-friendly principles are that guide urban planning and to what extent territories mirror them.

It is essential that the multidimensionality of people's experience of their environments be recognized at the risk of oversimplifying what are, by nature, rather complex questions. The major danger here is to overshadow the unmet needs of specific groups that might end up being obscured by policies more oriented to the needs of non-differentiated groups.

Our findings contribute to this debate and can be grouped as follows:

1. The WHO age-friendly cities project lists eight dimensions, encompassing several characteristics, of what an ideal city should be like to be considered an age-friendly environment. Some of the characteristics listed under each dimension relate to physical environments and yet others to social characteristics (WHO, 2007). The results from the Portuguese survey *cIDADES* suggest that more detailed groupings can be considered due to the multidimensional nature of each of the eight dimensions identified in the WHO checklist in order to better depict the detailed features of each city in Portugal. These more detailed groupings could be thought of in a two-dimensional space that opposes social dimensions to material dimensions and group-targeted approaches to universal standardized approaches.
2. The developed and applied multidimensional approach used with the Portuguese case points to some comprehensiveness and integration when planning for age-friendliness in Portuguese cities. However, the analysis along a two-dimensional space that contrasts social and physical environments and that pits group-targeted programs against universal policies signals a continuing mainstream approach to age-friendliness that favors physical environments and that draws on more general policies. This, however, comes together with a spillover effect that suggests that once the topic of age-friendliness is introduced in the agenda of urban planning and urban management at the local level, it is likely to increase sensitivity towards the subject in all of the domains of policy design and implementation.

Acknowledgments

We gratefully acknowledge the *Direção Geral de Saúde* and *Fundação Calouste Gulbenkian* grants as well as the *Fundação para a Ciência e a Tecnologia* grant (Project PEst-OE/SADG/UI0727/2014).

REFERENCES

Buffel, T., Phillipson, C., & Scharf, T. (2012). Ageing in urban environments: Developing "age-friendly" cities. *Critical Social Policy*, *32*, 597–617.

Burton, E., & Mitchell, L. (2006). *Inclusive urban design: Streets for life*. Oxford: Architectural Press.

Dijkstra, L., & Poelman, H. (2012). *Cities in Europe. The new OECD-EU definition*. Brussels: European Commission Regional Focus Report, RF 01/2012.

Gaspar, J. (1987). Portugal: os próximos 20 anos. In J. Gaspar (Ed.), *Ocupação e Organização do Espaço—Retrospectiva e Tendências* (Vol. I, pp. 5–122). Lisboa: Fundação Calouste Gulbenkian.

George, D., & Mallery, P. (2003). *SPSS for windows step by step: A simple guide and reference.11.0 update* (4th ed.). Boston, MA: Allyn & Bacon.

Green, G. (2012). Age-friendly cities of Europe. *Journal of Urban Health: Bulletin of the New York Academy of Medicine*, *90*, 116–128.

Harvey, D. (2008, September–October). The capitalist city. *New Left Review*, *53*, 23–42.

INE. (2012). *Censos 2011 Resultados Definitivos—Portugal*. Lisboa: Instituto Nacional de Estatística.

Lui, C., Everingham, J., Warburton, J., Cuthill, M., & Bartlett, H. (2009). What makes a community age-friendly: A review of international literature. *Australasian Journal on Ageing*, *28*(3), 116–121.

UN. (2012). *Population ageing and development*. New York: United Nations, Department of Economic and Social Affairs.

Verté, D. (2012). *Assessing age friendliness in cities*. Eurocities Social Affairs Forum. Brussels: Eurocities.

World Health Organization (WHO). (2007). *Global age-friendly cities: A guide*. Geneva, Switzerland: World Health Organization.

Appendix

Age-Friendly Cities and the WHO Checklist: Lessons from a Portuguese Survey

Alexandra Lopes, Teresa Pinto, and Rute Lemos

Methodological Considerations of the cIDADES Survey

CIDADES: THE SURVEY

cIDADES was launched in 2010 and involved a complex, mixed-method research design. The goal of the research project was to collect input from an array of social actors, ranging from academic experts and policymakers to ordinary citizens, in order to understand how different agents and stakeholders experience urban environments and how they assess the age-friendliness of the cities where they live and work. Local agents ranging from local authorities to NGOs were actively involved in the project at all stages. One of the tasks of *cIDADES* involved a survey of a national sample of individuals aged 55 and older, designed to understand how older populations experience their cities and to examine the current state of urban environments from the perspective of their older residents.

The questionnaire used for the survey was based on the WHO checklist and was organized into eight sections mirroring the dimensions of the original checklist.[8] Each one of the features included in each of the eight dimensions of the checklist was translated into a sentence that describes the presence of each feature in the city. Survey participants were asked about how each sentence described what takes place in their cities. Individuals could answer "yes/no" to each sentence; a spontaneous "don't know" option was also available. Some of the features in the original checklist required more than one sentence since they included more than one condition. For example, when the checklist states, "Sidewalks are non-slip, are wide enough for wheelchairs and have dropped curbs to road level," the questionnaire would disaggregate the item into three separate statements: "Sidewalks are non-slip," "Sidewalks are wide enough for wheelchairs," and "Sidewalks have dropped curbs to road level."

The questionnaire was tested before the survey: both cognitive testing and pre-testing were carried out. The testing of the questionnaire involved conferencing with experts within 10 sessions carried out across the country to introduce the survey. Some trials with older people were carried out. A total of 25 individuals were asked to answer the questionnaire, and minor changes were introduced for clarification of words and/or sentences.

SAMPLING CONSIDERATIONS

The initial goal of the research team, as formally agreed with the funding institution and adjusting to the resources made available for the fieldwork, was to work with a sample of around 1,500 respondents.[9] The survey was not set to involve a pure randomized sampling design since there was no access to population databases. The first stage in the sample design resorted to a stratified random selection of municipalities to provide a representative sample of the territories from which local authorities would be contacted and invited to join the survey. We worked with the municipality-level units, selected following the stratification of the national territory according to NUTs II, population density, and aging rate. The strata for population density and aging rate were defined according to the official INE (National Statistics Office) criteria. The final sample included data from 142 municipalities across the country and proportionally covered the territories as stratified. This number represents roughly 46% of the total municipalities in the country. The distribution of sampled municipalities by stratification criteria is proportional to the weight of each stratum in the population.

The following map of Portugal represents the territorial distribution of the sampled municipalities.

Contrary to some of the initial expectations of the research team, which anticipated some problems arising from refusals to participate, what ended happening was exactly the opposite. There were hardly any refusals; further, there was an overwhelming interest in the survey on the part of local authorities across the country. Each municipality provided the resources for carrying out the survey locally and collected far more questionnaires than initially requested. The initial sample, planned for 1,500 cases, turned into a database with close to 9,000 individuals. The research team trained the interviewers and monitored the data collection for fieldwork quality.

Based on the overall structure of the population of each participating municipality, a set of combined quotas was defined to account for gender distribution, weight of those 55 and older in the total population of the municipality, and age groups. All municipalities received their quotas with the instructions to use them as a reference for the final composition of the local sample. In many cases, the quotas were not strictly followed, and the size of the final sample largely exceeded that requested. Data were collected with purposive sampling in the urban areas of each of the sampled municipalities.

The final sample comprises 8,658 individuals, with local samples ranging from 50 to 1,117. No weighting procedure was used given the non-randomized sampling design. Furthermore, the dataset was only to be used to test some assumptions about the performance of the checklist and to identify broad-ranging trends in urban planning in Portugal. We shall not be establishing any comparisons between cities, not only because the data do

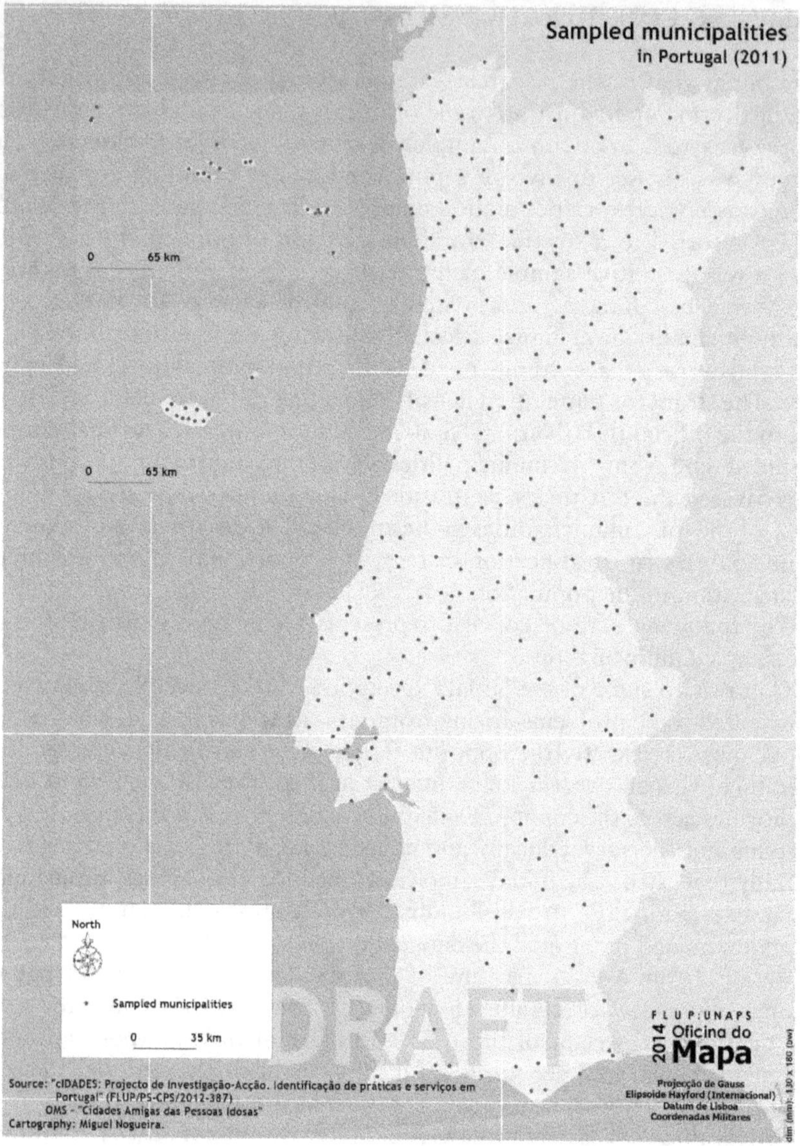

Figure 15.A.1 National Distribution of the Sampled Municipalities in cIDADES

not offer valid grounds for such comparisons, but also because the protocol included a clause of anonymity of municipalities in the discussion of results.

The population sample has the following sociodemographic makeup as displayed in Table 15.A.1. It reproduces overall the main features of the population of older adults in Portugal.[10]

Table 15.A.1 Sociodemographics for the Sample

Indicators		N	%	% of each demographic category in the Portuguese population aged 55+ according to the 2011 national census[a]
Age in intervals	55–60	1672	19.3	20.5
	60–65	1394	16.1	19.0
	65–70	1420	16.4	16.5
	70–75	1236	14.3	14.8
	75–80	958	11.1	13.1
	>80	885	10.2	16.1
Missing		1093	12.6	-
Sex	Male	3474	40.1	43.7
	Female	5080	58.7	56.3
Missing		104	1.2	
Marital status	Married/cohabiting	4870	56.2	66.2
	Divorced/separated	415	4.8	5.7
	Never married	544	6.3	6.7
	Widowed	2132	24.6	21.4
Missing		697	8.1	
Education level	Less than middle school	6361	73.5	78.8
	Middle school	345	4.0	8.2
	High school degree	593	6.8	5.7
	Higher education	202	2.3	7.3[b]
	College degree	250	2.9	
	Master degree	26	.3	
	PhD degree	7	.1	
Missing		874	10.1	
Status at work	Employed	901	10.4	18.1
	Unemployed	388	4.5	2.5
	Self-employed	257	3.0	[c]
	Volunteer	28	.3	[c]
	Retired	5004	57.8	69.6
	Other	555	6.4	9.8

(*Continued*)

Table 15.A.1 (Continued)

Indicators	N	%	% of each demographic category in the Portuguese population aged 55+ according to the 2011 national census[a]
Missing	1525	17.6	
Total	8658	100.0	
Final total after excluding missing cases across variables	8615	99.5	

Notes:

[a] Census data are last available for year 2011. Data available online at www.ine.pt.

[b] Education level in census data includes in the same category "higher education," "college degree," "Master's degree," and "PhD degree."

[c)]These two categories are merged into "other" in census data.

Non-response rate is not possible to compute given that there were no fieldwork logs kept by all interviewers. However, there are reasons to state that the non-response rate is marginal: face-to-face interviewing was the preferential mode of administration of questionnaires, and feedback from interviewers points to few cases of refusal.

"Don't know" was a spontaneous category of response that also ended up generating a marginal frequency. The share of "don't know" responses was between 0.7% and 1.5% across questions. Some tests were run to analyze for distribution bias in "don't know" responses: gender differences, place of inquiry, age group, and educational level were tested for correlation with "don't know" answers and no significant associations were identified. Also, no pattern was established in the distribution of "don't know" responses across individuals. After some consideration, although from a substantive point of view, "don't know" responses are different from an explicit statement of non-existence of the condition portrayed in the respective sentence in the questionnaire, it was considered appropriate to recode "don't know" responses into non-recognition of the condition. The argument to sustain this option was that not knowing about a specific condition for practical purposes means not having access or not enjoying that same condition.

The interpretation of results fails to account for the impact of territorial characteristics in the performance of the checklist, but some attempts to account for the stability of results at the local level suggest that that impact may be marginal.

ANALYSIS OF DATA

A principal components analysis was performed on the items of the checklist as phrased in the survey questionnaire and following the groupings of the original checklist. The first goal was to check for the one-dimensionality of each grouping of indicators as found in the WHO checklist. For all eight models, the results of both KMO indexes and Bartlett's tests of sphericity show a good fit (KMO ranging from a low of 0.803 to a high of 0.933). Criteria for the extraction of components revealed, cumulatively, an eigenvalue higher than 1 and more than two indicators in the dimension. Orthogonal rotations were used to facilitate the interpretation of the matrix of components and the assignment of each indicator to a component. Original variables were reverse-coded to guarantee that all items loaded positively in the extracted components. This was done to facilitate interpretation of loadings.

Answers of "don't know" were recoded as "no" or "yes" as applicable. Although one could argue that this over-represents the absence of items, for the purpose of this discussion, we were primarily interested in counting the awareness of specific features. For the purpose of measurement, a spontaneous "don't know" answer was regarded as a lack of awareness of the target condition.

The components were later used to compute summary indexes using simple additive computation. The new generated variables are numerical indexes that vary between 0 and 100. Internal consistency of items was assessed resorting to the Cronbach's alpha model, and indexes were computed for all components with an alpha higher than 0.6. We followed the rule of thumb as stated by George and Mallery (2003, p. 231) (=> .9 — Excellent, => .8 —Good, => .7 —Acceptable, => .6 —Questionable, = > .5 —Poor, and < .5 —Unacceptable) for deciding on the alpha value. This is an exploratory study and, while keeping in mind the need for some caution, those dimensions with an alpha of 0.6 were still analyzed.

The summary indexes were later analyzed resorting to a K-means cluster analysis to identify relatively homogeneous groups of cases (municipalities). There was no initial definition of cluster centers, and the analysis was run several times to account for case-ordering bias with the option running means active. Distances were computed using simple Euclidean distances. The analysis was run with iteration. Initial number of clusters was specified to three. The purpose was to test a grouping of municipalities that would be sorted out as low, average, and high scoring. The final cluster centers were then used to plot a radar-shape graphic representation. The objective was to display the relative position of each subgroup of municipalities across the input variables (indexes) used to obtain the cluster solution.

NOTES

1 For a detailed presentation of the *cIDADES* research project, please refer to the methodological appendix.
2 The old-age support ratio is defined as the ratio between the population aged 15 to 64 and the population aged 65 or older.
3 Law no. 11/82 —June 12, 1982.
4 See appendix.
5 For a detailed description of the procedures involved in computing the figures displayed in Table 15.1, please refer to the methodological appendix.
6 For each dimension we have computed a score using a simple additive process (weighting by factor scores did not show any significant difference, and the additive process provided a more intuitive interpretation). For internal consistency Cronbach's alphas below 0.6 were eliminated ("Traffic and roads," "Bus drivers' attitude"). Scores vary between 0 and 1. Values closer to 1 indicate a higher level of awareness by the population about the availability of the domain in question.
7 See appendix.
8 The pdf with the complete questionnaire can be accessed from the author (aslopes@letras.up.pt). The questionnaire is available only in the Portuguese language.
9 The sample size was set for a confidence level of 99.7% (± 3σ) and a marginal error of ±4%.
10 Although all indicators seem to show relatively high numbers of missing cases, of all missing cases, only 43 were missing in all sociodemographic indicators. These individuals were excluded from the sample. Their distribution by municipality was random. The remaining missing cases, by sociodemographic variable, were tested for association with the municipality and among themselves with no statistically significant association identified.

Contributors

M. Scott Ball, M.Arch., is Principal, Commons Planning, Atlanta, Georgia, USA.

John R. Beard, MBBS, PhD, is Director of the Department of Ageing and Life Course at the World Health Organization in Geneva, Switzerland.

Marie Beaulieu, PhD, is Full Professor and Researcher at University of Sherbrooke, School of Social Work, Research Centre on Aging, Quebec, Canada.

Tine Buffel, PhD, is Postdoctoral Researcher, Department of Adult Educational Sciences, Vrije Universiteit Brussel, Brussels, Belgium, and the Cathie Marsh Centre for Census and Survey Research, School of Social Sciences, University of Manchester, Manchester, United Kingdom.

Francis G. Caro, PhD, is a sociologist and gerontologist. He is a professor emeritus of Gerontology at the University of Massachusetts Boston. He is the editor of the *Journal of Aging & Social Policy*. As a volunteer, he is a co-chair of an age-friendly city initiative in Brookline Massachusetts where he lives.

Mihye Cho, PhD, is Assistant Professor at International Design Center, Singapore University of Technology and Design, Singapore.

Keng Hua Chong, PhD, is Assistant Professor at International Design Center, Singapore University of Technology and Design, Singapore.

Joan K. Davitt, PhD, is Associate Professor, School of Social Work, University of Maryland, Baltimore, Maryland, USA.

Liesbeth De Donder, PhD, is Professor in Adult Educational Sciences, Vrije Universiteit Brussel, Brussels, Belgium.

Nico De Witte, PhD, is Lecturer in Adult Educational Sciences, Vrije Universiteit Brussel, Brussels, and Lecturer in Health Sciences, University College Ghent, Ghent, Belgium.

Alan DeLaTorre, PhD, is Research Associate, Portland State University, Institute on Aging, Portland, Oregon, USA.

Sarah Dury, MA, is a Doctoral Student in Adult Educational Sciences, Vrije Universiteit Brussel, Brussels, Belgium.

Billy Fields, PhD, is Assistant Professor of Political Science, Texas State University, San Marcos, TX, USA.

Kelly G. Fitzgerald, MPA, PhD, received a gerontology doctorate from the University of Massachusetts Boston. She has a master's degree in public administration and gerontology. She resides in Switzerland and is a senior scientist and adjunct graduate faculty member at Western Kentucky University's Center for Gerontology. She has worked with the World Health Organization and provided support to help cities apply to join the WHO Global Network of Age-Friendly Cities and Communities, link cities with other cities, and identify ways in which cities can become more age-friendly. She is the Vice-Chair of the United Nations NGO Committee on Ageing in Geneva, Switzerland, where she represents the International Federation on Ageing.

Suzanne Garon, PhD, is Full Professor and Researcher at University of Sherbrooke, School of Social Work, Research Centre on Aging, Quebec, Canada.

Robert Geary, BA, Managing Editor, has been Managing Editor for the *Journal of Aging & Social Policy* for the past 15 years.

Allen Glicksman, PhD, is Director of Research and Evaluation, Philadelphia Corporation for Aging, Philadelphia, Pennsylvania, USA.

Carrie L. Graham, PhD, is Assistant Director of Research, Health Research for Action, School of Public Health, University of California at Berkeley, Berkeley, California, USA.

Emily A. Greenfield, PhD, is Assistant Professor, School of Social Work, Rutgers, The State University of New Jersey, New Brunswick, New Jersey, USA.

George C. Homsy, PhD, is Assistant Professor, Department of Public Administration, Binghamton University, Binghamton, New York, USA.

Zheng Jia, MCP, is Researcher at International Design Center, Singapore University of Technology and Design, Singapore.

Maureen Kelly, MSW, is Business & Community Liaison, Aging Services Division, Atlanta Regional Commission, Atlanta, Georgia, USA.

Laura Keyes, AICP, MS, is Doctoral Student, Department of Public Administration, University of North Texas, Denton, Texas, USA.

Andréanne Laliberté is a MSW Candidate at University of Sherbrooke, School of Social Work, Research Centre on Aging, Quebec, Canada.

Kathryn Lawler, MPP, is Manager of Aging and Health Resources, Atlanta Regional Commission, Atlanta, Georgia, USA.

Amanda J. Lehning, PhD, is Assistant Professor, School of Social Work, University of Maryland, Baltimore, Maryland, USA.

Rute Lemos is Doctoral Student, Sociology, University of Porto, Porto, Portugal.

Debbie Loo, M.Arch., is Researcher at International Design Center, Singapore University of Technology and Design, Singapore.

Alexandra Lopes, PhD, is Professor of Sociology, University of Porto, Porto, Portugal.

Tyrone Manegdeg, is a graduate student at the College of Architecture, Georgia Institute of Technology, Atlanta, Georgia, USA.

Cheryl Mayerik is Communications Coordinator, Atlanta Regional Commission, Atlanta, Georgia, USA.

Jennifer McArthur, MSc, is Research Assistant, Department of Community Health Sciences and Centre on Aging, University of Manitoba, Winnipeg, Manitoba, Canada.

Paul McGarry, MA, is Senior Strategy Manager, Public Health Manchester, and Senior Strategy Manager, Manchester City Council, Manchester, United Kingdom.

Verena H. Menec, PhD, is Professor, Department of Community Health Sciences and Director, Centre on Aging, University of Manitoba, Winnipeg, Manitoba, Canada.

Gulnara A. Minnigaleeva, PhD, is Research fellow, Associate Professor, Centre for Studies of Civil Society and Nonprofit Sector, National Research University Higher School of Economics; CEO, Organization of Retired Persons "Wisdom Ripening," Bashkortostan, Russian Federation.

Margaret B. Neal, PhD, is Director and Professor, Portland State University, Institute on Aging, Portland, Oregon, USA.

Sheila Novek, MSc, is Research Coordinator, Centre on Aging, University of Manitoba, Winnipeg, Manitoba, Canada.

Hugh O'Connor, BBS, MBS, is National Program Manager, Age-Friendly Cities and Counties Program, Dublin, Ireland.

Mario Paris is a Doctoral Candidate at University of Sherbrooke, School of Social Work, Research Centre on Aging, Quebec, Canada.

Deborah R. Phillips, PhD, CPM, is Adjunct Professor, Housing and Consumer Economics, University of Georgia, Athens, and Adjunct Professor, Building Construction, Georgia Institute of Technology, Atlanta, Georgia, USA.

Chris Phillipson, PhD, is Professor of Sociology and Social Gerontology, Cathie Marsh Centre for Census and Survey Research, School of Social Sciences, University of Manchester, Manchester, United Kingdom.

Teresa Pinto is President of Associação VIDA—VIDA Association (Intergenerational Dissemination and Active Development), Lourosa, Portugal.

Lauren Ring, BA, is Planning Associate, Philadelphia Corporation for Aging, Philadelphia, Pennsylvania, USA.

Kristen N. Robinson, PhD, is Senior Research Scientist, Social & Scientific Systems, Inc., Silver Spring, Maryland, USA.

Andrew E. Scharlach, PhD, is Kleiner Professor of Aging, School of Social Welfare, University of California at Berkeley, Berkeley, California, USA.

Sinead Shannon, MA, is MSc Research Manager, Age-Friendly Ireland, Dublin, Ireland.

An-Sofie Smetcoren, MA, is a Doctoral Student in Adult Educational Sciences, Vrije Universiteit Brussel, Brussels, Belgium.

Evelina Sterling, PhD, MPH, MCHES, is Public Health Consultant, Atlanta Regional Commission, Atlanta, Georgia, USA.

Kathleen E. Sykes, MA, is Senior Advisor for Aging and Sustainability, Office of Research and Development, United States Environmental Protection Agency, Washington, District of Columbia, USA.

Grace Trimble, BS, MA, is Communications Coordinator, Atlanta Regional Commission, Atlanta, Georgia, USA.

Jason Tudor, MURP, is Director of Outreach and Education, AARP Louisiana, New Orleans, Louisiana, USA.

Anne Veil, MSW, is research coordinator at University of Sherbrooke, School of Social Work, Research Centre on Aging, Quebec, Canada.

Dominique Verté, PhD, is Professor of Adult Educational Sciences, Vrije Universiteit Brussel, Brussels, Belgium.

Dawn Veselyuk, BSW, is Project Coordinator, Centre on Aging, University of Manitoba, Winnipeg, Manitoba, Canada.

Mildred E. Warner, PhD, is Professor, Department of City and Regional Planning, Cornell University, Ithaca, New York, USA.

Lisa Warth, PhD, is a Technical Officer at the Department of Ageing and Life Course. She coordinates the WHO Global Network of Age-Friendly Cities and Communities connecting hundreds of cities and communities worldwide that are committed to becoming more age-friendly.

For Product Safety Concerns and Information please contact our EU representative GPSR@taylorandfrancis.com
Taylor & Francis Verlag GmbH, Kaufingerstraße 24, 80331 München, Germany

www.ingramcontent.com/pod-product-compliance
Lightning Source LLC
LaVergne TN
LVHW011115300926
849160LV00008B/397

* 9 7 8 1 1 3 8 5 4 6 3 6 3 *